Usharani Rathinam Krishnaswamy
Gustavo Henrique Ribeiro Da Silva

AF587396

Microbial Biodegradation, Bioremediation and Ozonation Technology for Sustainable Organophosphate Remediation *(Methylparathion-Insecticide)*

www.novapublishers.com

Copyright © 2025 by Nova Science Publishers, Inc.
DOI: https://doi.org/10.52305/YGDV4272

All rights reserved. No part of this book may be reproduced, stored in a retrieval system or transmitted in any form or by any means: electronic, electrostatic, magnetic, tape, mechanical photocopying, recording or otherwise without the written permission of the Publisher.

We have partnered with Copyright Clearance Center to make it easy for you to obtain permissions to reuse content from this publication. Simply navigate to this publication's page on Nova's website and locate the "Get Permission" button below the title description. This button is linked directly to the title's permission page on copyright.com. Alternatively, you can visit copyright.com and search by title, ISBN, or ISSN.

For further questions about using the service on copyright.com, please contact:
Copyright Clearance Center
Phone: +1-(978) 750-8400 Fax: +1-(978) 750-4470 E-mail: info@copyright.com.

NOTICE TO THE READER

The Publisher has taken reasonable care in the preparation of this book, but makes no expressed or implied warranty of any kind and assumes no responsibility for any errors or omissions. No liability is assumed for incidental or consequential damages in connection with or arising out of information contained in this book. The Publisher shall not be liable for any special, consequential, or exemplary damages resulting, in whole or in part, from the readers' use of, or reliance upon, this material. Any parts of this book based on government reports are so indicated and copyright is claimed for those parts to the extent applicable to compilations of such works.

Independent verification should be sought for any data, advice or recommendations contained in this book. In addition, no responsibility is assumed by the Publisher for any injury and/or damage to persons or property arising from any methods, products, instructions, ideas or otherwise contained in this publication.

The Publisher assumes no responsibility for any statements of fact or opinion expressed in the published contents.

This publication is designed to provide accurate and authoritative information with regard to the subject matter covered herein. It is sold with the clear understanding that the Publisher is not engaged in rendering legal or any other professional services. If legal or any other expert assistance is required, the services of a competent person should be sought. FROM A DECLARATION OF PARTICIPANTS JOINTLY ADOPTED BY A COMMITTEE OF THE AMERICAN BAR ASSOCIATION AND A COMMITTEE OF PUBLISHERS.

Additional color graphics may be available in the e-book version of this book.

Library of Congress Cataloging-in-Publication Data

ISBN: 979-8-89530-685-7 (Softcover)
ISBN: 979-8-89530-546-1 (eBook)

Published by Nova Science Publishers, Inc. † New York

Above all, I express deep gratitude to my beloved family, my beloved parents, sisters, and brothers for their constant and kind support, which has greatly encouraged me every moment to give my best towards this work. They have been with me always in my endeavors. I owe a sense of gratitude to the Almighty for his blessings to complete this work successfully.

Beyond everything, I dedicate this work and book to my beloved mother, Mrs. Rathinam Krishnaswamy, and father, Mr. Krishnaswamy C. P.

Dr. Usharani Rathinam Krishnaswamy

Contents

List of Tables

List of Figures

Preface

This book is based on my experience with PhD Research and Research Projects, as well as with teaching in the following subjects: Environmental Science, Ecology, Energy and Environment, Instrumental Methods and Environmental Microbiology, and Bioremediation to MSc students (Environmental Sciences), and BSc (Environmental Sciences, Microbiology) courses of Indian universities and colleges.

Bioremediation is a branch of biotechnology that employs the use of living organisms, like microbes and bacteria, in the removal of xenobiotic contaminants, pollutants, and toxins from soil, water, and other environments. This book has been written in a clear, lucid manner to cover theoretical and applied aspects of bioremediation of phosphate and nitrate nutrient-rich wastewater. It will help postgraduate and graduate students, research scholars, and teachers of environmental sciences, microbiology, and biotechnology. I believe this book's contents are easy to understand without needing any external help. The research articles are helpful for future analysis of this significant field of study to the researcher's students develop their field of interest, focusing on sustainable development and management of the environment. Several references, books, articles, and research papers have helped in preparing this book.

The objective of the book was to examine the efficient Biodegradation, Bioreduction, Bioremoval, and Biotreatment of organophosphate pesticides including Parathion, Methylparathion, from contaminated environments using microbial bioremediation and ozonation processes and draw a valid conclusion.

The book describes the Organophosphate Pesticides-Methylparathion contaminated environment and its wastewater treatment methods by Sustainable Technologies using Microorganism, Bioremediation Methods, Techniques, Ozonation Process, and Technology. It deals with 14 chapters including Introduction and Conclusion along with 12 figures and 7 Tables.

With the above objectives, the following book chapters are framed and analyzed from the review work of research papers. The chapters in this book include the following line: Chapter 1 offers an introduction on the xenobiotics, organophosphate pesticides, and methylparation usages in India at the national level as well as global scenarios. Chapter 2, 3 and 4 deals with an overview of organophosphate pesticides, their major sources, effects, and pollution. Chapters 5, 6, 7 and 8 describe organophosphate, methylparathion pesticides, chemistry, structures, features, identity, chemical and physical properties, and their effects on health and the environment. Chapter 6 deals with the toxicity of these compounds, and Chapter 9, discusses disposable methods for treating or removing these pollutants, and compounds from the environment it includes water and soil ecosystems, and their networks. Chapter 10 deals mainly with microbial treatment, Biodegradation Process, bio-transformation, bio-conversion, bio-reduction, bio-oxidation, and Bioremediation Technology of these xenobiotic pesticides. Also describes the conversion pathways, metabolites of products of primary and secondary intermediates formation, end products, etc. It also stated the key enzymes of microbial sources or origin responsible for the non-toxic end products and biodegradations. Chapter 11 describes the various Environmental factors responsible for the biodegradation process of these xenobiotic pesticides, and its optimized conditions responsible for complete or partial biodegradation. Chapter 12 deals with the Ozonation process and Technology used for the removal and remediation of these pesticide compounds in wastewater treatment, and Chapter 13, states the Process Optimization, Response surface methodology, and key strategies to optimize and reduce these pollutants from contaminated or polluted Environments. Finally, Chapter 14 deals with the conclusion of these disposable methods, sustainable technologies using bioremediation and ozonation techniques, strategies, process optimization, its sustainable management of organophosphate pesticide, methylparathion from the ecosystem.

Acknowledgments

I take the opportunity to express my sincere thanks to Professor Dr. Lakshmanaperumalsamy, Emeritus Professor, Department of Environmental Sciences, Bharathiar University, India, whose encouragement has driven me to prepare this book. I wish to express my sincere thanks to all my directors and professors in DRDO-BU Center for Life Sciences, Bharathiar University, Coimbatore, India, for providing Research Fellowships SRF, JRF to carry out my research work. My appreciation goes to NOVA Science Publishers for their kind support.

Dr. Usharani Rathinam Krishnaswamy

UN Sustainable Development Goals

Source: UN 2019 https://sdgs.un.org/2030agenda.

The Sustainable Development Goals and targets recognize the critical role that a healthy environment can play in addressing current challenges including poverty, climate change, food and water security. It recognizes the urgent need to protect the natural world, both for its own sake, and to meet the needs of more than 9 billion people by 2050.

Goal 1. End poverty in all its forms everywhere.
Goal 2. End hunger, achieve food security and improved nutrition, and promote sustainable agriculture.
Goal 3. Ensure healthy lives and promote well-being for all at all ages.
Goal 4. Ensure inclusive and equitable quality education and promote lifelong learning opportunities for all.
Goal 5. Achieve gender equality and empower all women and girls
Goal 6. Ensure availability and sustainable management of water and sanitation for all.
Goal 7. Ensure access to affordable, reliable, sustainable, and modern energy for all.
Goal 8. Promote sustained, inclusive and sustainable economic growth, full and productive employment and decent work for all.
Goal 9. Build resilient infrastructure, promote inclusive and sustainable industrialization and foster innovation.
Goal 10. Reduce inequality within and among countries.
Goal 11. Make cities and human settlements inclusive, safe, resilient and sustainable.
Goal 12. Ensure sustainable consumption and production patterns.
Goal 13. Take urgent action to combat climate change and its impacts.
Goal 14. Conserve and sustainably use the oceans, seas and marine resources for sustainable development.
Goal 15. Protect, restore and promote sustainable use of terrestrial ecosystems, sustainably manage forests, combat desertification, halt and reverse land degradation and halt biodiversity loss.
Goal 16. Promote peaceful and inclusive societies for sustainable development, provide access to justice for all and build effective, accountable and inclusive institutions at all levels.
Goal 17. Strengthen the means of implementation and revitalize the global partnership for sustainable development.

Introduction

Organophosphate Phosphate pollutants show the way to amplify the growth of toxicity and bioaccumulation of toxicants in the environment. It has become the most serious environmental problem it causes groundwater and surface water pollution and decreases the dissolved oxygen concentration of water which results in harmful effects to all living organisms and their ecosystems. This study was focused on waste stabilization and pesticide removal and examined the biodegradation, and bioremoval potential of organophosphate by selective potential microorganisms, bacteria (*Pseudomonas spp, Bacillus spp)*, fungi (*Fusarium* spp), *a*lgae, cyanobacteria, and Ozonation process using ozone molecule from pesticide-contaminated water and soil according to the variance of residence time, temperature, pH and concentrations of organophosphate xenobiotic pollutants. The study showed potential biodegradation, bioremoval, and biotreatment processes for the purification of water and effectively removed the pesticide methylparathion concentrations using cost-effective solutions and eco-friendly Ecotechnology. There it improves the quality of different wastewater and is used for various purposes for sustainable applications include agriculture and aquaculture activities. In brief, the amount of *biomass* is increased considerably in wastewater containing high concentrations of nutrients under controlled and optimized bioprocess conditions of different biological variables and their corresponding parameters. Hence, the combined effect of both microbial and Ozonation processes was used potentially for effective biodegradation, bioremoval, and biotreatment process for organophosphate (methylparathion) rich wastewater under the optimized variables, which act as a cost-effective solution and eco-friendly Ecotechnology. There it was also possible to improve the quality of different pesticide-contaminated wastewater and treated wastewater which can be used for agriculture and aquaculture activities which in turn is helpful for sustainable development and management of wastewater.

Keywords: xenobiotics, organophosphate, pesticides, biodegradation, microbial bioremediation, ozonation process, sustainable development and management

Chapter 1

Organophosphate Xenobiotics

During the last century, advances in synthetic chemistry have given chemists the ability to make numerous novel compounds, some of which are xenobiotic (Xu et al. 1999). The release of xenobiotic compounds into the environment and the problem of toxic waste disposal have become enormous due to the proliferation of these xenobiotic compounds for use as pesticides, solvents, explosives, refrigerants, and dyes in industrial, urban, and agricultural applications (Doung et al. 1997). Many xenobiotic compounds, particularly those used as insecticides, are toxic (Xu et al. 1999). Even though many insecticides degrade rapidly in the soil, they can be potentially hazardous as a consequence of accidental spills, runoff from applications in agricultural areas, and discharge from pesticide containers and waste disposal systems (Hertel, 1993; Rani and Lalithakumari, 1994; Shimazu et al. 2001; Zhongli et al. 2001).

Organophosphate compounds represent the largest group of chemical insecticides used in plant protection throughout the world. In the United States, Organophosphorus represents about half of the total insecticides used, with annual applications of over 75 million pounds (Bravo et al. 2002). Because of the rapid growth in industrial and agricultural chemical usage, vast quantities of soil and groundwater have been contaminated with hazardous compounds (Chauhan et al. 2000). Environmental contamination by toxic xenobiotic chemicals, arising mainly from agricultural and industrial sources, and the consequences on food quality and human health are very serious worldwide problems. The extensive use of pesticides (fungicides, herbicides, insecticides) in various parts of the world for years without any control was responsible for a dramatic increase in pollution levels in soil and water. Moreover, fungicides and herbicides were often randomly used, resulting in inefficiency in soil and ecosystem contamination.

Organophosphorus is used worldwide as a pesticide and petroleum additive. Organophosphorus compounds share the major portion of the pesticide market globally. Owing to the large-scale use of organophosphorus compounds, contaminations of soil and water systems have been reported from all parts of the world. Organophosphorus compounds possess very high mammalian toxicity and therefore early detection and subsequent decontamination and detoxification of the polluted environment is essential. Additionally, about 200,000 tons of extremely toxic organophosphate chemical warfare agents are required to be destroyed by 2007 under the Chemical Warfare Convention of 1993. Many xenobiotic compounds, particularly those used as insecticides, are toxic (Xu et al. 1999).

Recently, endocrine-disrupting chemicals (EDCs) have become more widespread in the environment and have been found to deteriorate the generative function of some living species on the earth (Harries et al. 2000). Chemical substances that can interfere with the normal functioning of the endocrine system have been termed Endocrine endocrine-disrupting chemicals (EDCs) (Keith et al. 1997). Endocrine-disrupting chemicals (EDCs) have been shown to produce changes in the endocrine system of organisms that may lead to an increase in cancers and abnormalities in reproductive structure and function. Recent research has highlighted the existence of hormonally active compounds in sewage and industrial effluents and their potential for recycling back into the environment including drinking water supplied through point and non-point sources. On the other hand, surplus pesticides, even though within their expiration limits, will become a burden when their future use is prohibited by legislation due to toxicological or environmental concerns. The disposal of EDCs such as pesticides, fungicides, and insecticides can cause serious environmental problems due to the chemical toxicity of the active ingredients in the formulations and sometimes, of their decomposition products.

Pesticides are substances that are used to control, destroy, repel, or attract pests to minimize their detrimental effects. Pesticides are used in many situations such as livestock farming, cropping, horticulture, forestry, home gardening, homes, hospitals, kitchens, roadside, recreational, and industrial areas. By 1991, there were approximately 23,400 pesticide products registered with the U.S. Environmental Protection Agency (EPA). The agriculture industry used 77%, industrial, commercial, and government organizations used 12% and private households used the remaining 11%. In studies of laboratory rodents exposed to extremely high doses, about half of all man-made pesticides are found to be carcinogenic. On a list of most common POISONS, pesticides were found to be number two. According to the Environmental Protection Agency (EPA), 60% of herbicides, 90% of fungicides, and 30% of insecticides are known to be carcinogenic (EPA, U.S., Pesticides Industry Sales and Usage: 1996 and 1997 and EPA, U.S., Pesticides: Topical & Chemical Fact Sheets, 2008). In 1997, 1.2 billion pounds of pesticides were used. The agriculture industry used 77%, industrial, commercial, and government organizations used 12% and private households used the remaining 11%. (EPA, 2000).

The literature review provides basic scientific information about the classification of pesticides in use and pesticide usage patterns in India and the world. The review shows the current scenario of pesticide usage patterns in Indian agriculture.

Table 1a. Pesticide classification based on their toxicity

WHO Class		LD_{50} for rats (mg/kg of body weight)	
		Oral	Dermal
Class-I_a	Extremely Hazardous	Less than 5	Less than 5
Class-I_b	Highly Hazardous	5 to 50	5 to 200
Class-II	Moderately Hazardous	50 to 2000	200 to 2000
Class-III	Slightly Hazardous	Over 2000	Over 2000
Class-V	Unlikely to present acute hazard	5000 or higher	

1.1. Pesticide Classification Based on the Following Criteria

1. Classification of pesticides according to their toxicity
2. Classification of Pesticides according to Chemical Composition:
3. Classification of pesticides based on the pest organism they kill and the pesticide's functionality (Use)
4. Classification of pesticides based on Mode of Entry: Pesticide modes of entry refer to the various ways pesticides come into touch with or enter the target.
5. Classification of pesticides by mode of action
6. Classification based on sources of origin

This is the most popular and useful way of pesticide classification based on chemical makeup. Pesticides such as insecticides, fungicides, herbicides, and rodenticides are also classed based on their chemical compositions, as shown below:

- *Insecticides:* Insecticides are classed chemically as Carbamates (Carbaryl), Organochlorine (Endosulfan), Organophosphorus (Monocrotophos), Pyrethroids (permethrin), Neonicotinoids

(Imidacloprid), various pesticides such as Spinosyns (Spinosad), Benzolureas (diflubenzuron), Antibiotics (abamectin).

- *Fungicides:* Fungicides are categorized as aliphatic nitrogen fungicides (dodine), amide fungicides (carpropamid), aromatic fungicides (chlorothalonil), dicarboximide fungicides (famoxadone), dinitrophenol fungicides (dinocap), and others.
- *Herbicides:* Herbicides include anilide herbicides (flufenacet), phenoxyacetic herbicides (2, 4-D), quaternary ammonium herbicides (Paraquat), chlorotriazine herbicides (atrazine), sulfonylurea herbicides (chlorimuron), and others.
- *Rodenticides:* Rodenticides are classed as inorganic rodenticides (Zinc phosphide, Aluminium Phosphide) or organic coumarin rodenticides (bromadiolone, coumatetralyl).

Chapter 2

Global and National Status

The organophosphate compounds are most commonly associated with serious human toxicity, accounting for more than 80% of pesticide-related hospitalizations (Haddad and Winchester, 1983). In contrast to the past, when chlorinated hydrocarbon compounds such as DDT were commonly used, organophosphate insecticides have become increasingly popular for both agricultural and home use because their unstable chemical structure leads to rapid hydrolysis and little long-term accumulation in the environment (Namba, 1971). This widespread use, however, has resulted in increased numbers of human poisonings. Through the 1970s, the Environmental Protection Agency estimated that 3,000 hospitalizations per year were required for insecticide poisoning in the United States, with a fatality rate of 50% in the pediatric age group and 10% in adults (EPA, 1979). In 1983 data from the American Association of Poison Control Centres indicated that the national incidence of insecticide exposures was 77,000, of which 33,000 involved organophosphates (Veltri and Litovitz, 1983). The continued use of such chemicals will likely increase these statistics in the future. The first global estimates of the extent of pesticide poisoning were published in 1990 by the World Health Organisation (WHO, 1990). Based on extrapolations from limited data, it was estimated that 3 million cases of pesticide poisonings occurred worldwide annually with 220,000 deaths, the majority intentional.

The WHO estimates, based on 2001 data, that 849,000 people die globally from self-harm each year (WHO, 2002). How many of these cases are a result of poisoning with pesticides is not known. However, poisoning is the most commonest form of fatal self-harm in rural Asia, accounting for over 60% of all deaths (Somasundaram and Rajadurai, 1995; Phillips and Zhang, 2002; Joseph, 2003), and is of far greater importance than hanging, and other physical forms of self-harm. Furthermore, a review of poisoning studies reveals that pesticides are the commonest means of self-poisoning in many rural areas and are associated with a high mortality rate (Eddleston, 2000). A recent national survey in Bangladesh showed that 14% of all deaths (3971 of 28,998) of women between 10 and 50 years of age were due to self-poisoning, the majority with pesticides (Yusuf, 2000). The problem is particularly severe in Sri Lanka (Berger, 1988) where pesticide poisoning was the most common cause of hospital death in six rural districts in 1995. In many countries, the widespread availability of acutely toxic pesticides used in agriculture has made a selection of pesticides as the agents of choice for self-harm well-known to both healthcare workers and public health authorities (Kasilo et al, 1991, Daisley and Hutchinson, 1998).

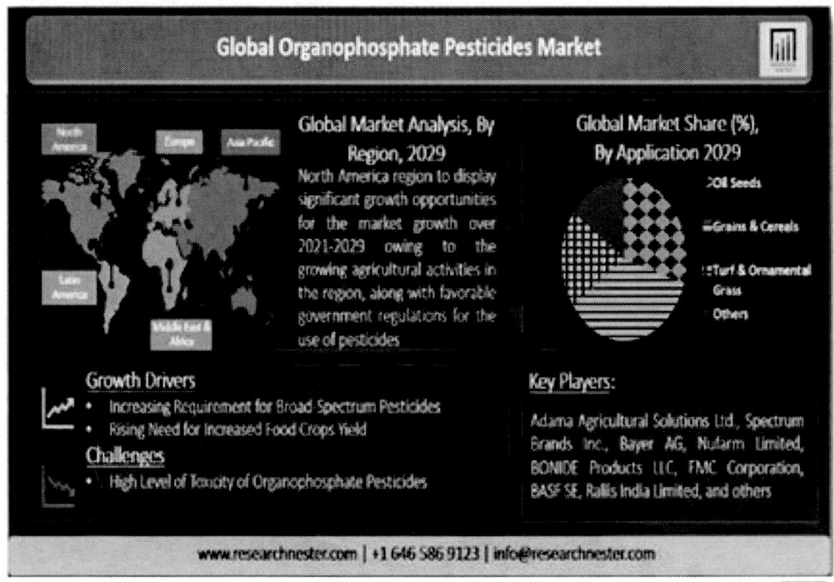

Global Organophosphate Pesticides Market Share (in %), Segmentation by Region, 2029

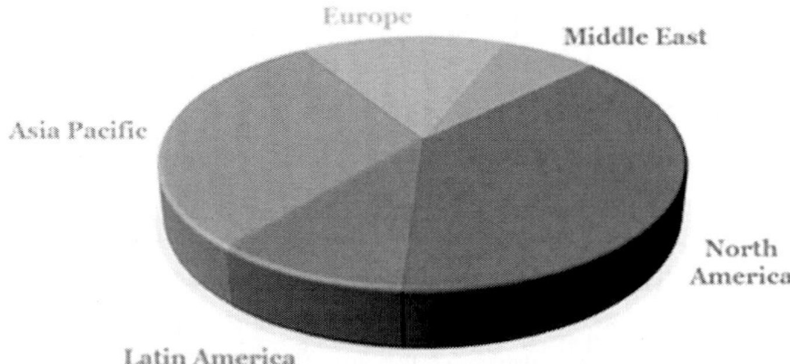

Source: Research Nester.

Figure 1a. Global Organophosphate Pesticides Market.

The importance of pesticides in India can be understood from the fact that agriculture is a major component of the Indian economy: It contributes 22% of the nation's GDP and is the livelihood of nearly 70% of the country's workforce. Globally, due to consolidation in the agrochemical industry, the top five multinational companies control almost 60% of the market. In India, the industry is very fragmented, with about 30 to 40 large manufacturers and about 400 formulators. The use pattern is skewed towards insecticides, which accounted for 67% of the total pesticide consumption in 2006 (Figure 1a). The potential adverse impact on human health from exposure to pesticides is likely to be higher in countries like India due to easy availability of highly hazardous products, and low risk awareness, especially among children and women. Overexposure to pesticides can occur before spraying– because of easy access for children, lack of adequate labeling, and during mixing – during spraying and after spraying operations. Spray operators and bystanders can be affected. Having cheap and easily available highly hazardous pesticides at hand increases the incidence of intentional pesticide poisonings (WHO, 2009). The effective number of cases of pesticide poisoning occurring in India annually has been estimated by Ravi et al. (2007) to be up to 76,000, much higher than the figure of NCRB. Furthermore, Gunell et al. (2007) calculate that the number of intentional cases alone reaches some 126,000 cases annually (Ravi et al. 2007). Figure 1a represents the global organophosphate pesticides market in worldwide consumption data.

Chapter 3

The Indian Scenario

India is primarily an agricultural-based country with more than 60-70% of its population dependent on agriculture As India's population is increasing day by day there is more pressure on the annual food grain production and on minimizing crop losses (Shroff, 2000). India's fast-growing population is projected to cross 1.3 billion by 2020 (US Census Bureau, Report of the National Intelligence Council (2020), www.foia.cia.gov/2020/2020.pdf). To feed and clot this population from almost exhausted cultivatable land and fast-depleting water resources would be a great concern and challenge to the Indian farmers. The increase in agricultural output and the phenomenal progress that has been made is due to the application of improved technology, and increased use of important agricultural inputs like fertilizers, hybrid seeds, and pesticides. About 30% of agricultural produce is lost due to pests and crops worth rupees 60,000 crores are just eaten away by pests (NABARD's Report, www.nabard.org). Therefore extensive use of pesticides is inevitable. Pesticides provide a sure cover to the farmer in protecting his investment in seeds, fertilizers, irrigation, and hard labor from insects and pests. Pesticides constitute the key control strategy for crop disease and pest management and have been making significant contributions in India towards improving crop yield. Hence agrochemicals like insecticides, fungicides, pesticides, and herbicides have increasing demand for the protection of crops in one way or the other (Anonymous 2000). Thus, large numbers of pesticides are emerging day by day. India is presently the second-largest manufacturer of basic pesticides in Asia. It ranks 12th globally (Mathur, 1999, Gupta, 2004). The worldwide consumption of pesticides is about two million tonnes per year, of which 24% is consumed in the USA alone, 45% in Europe, and 31% in the rest of the world. India's share is just 3.75%. The usage of pesticides in India is only 0.5 kg/ha, while in Korea and Japan, it is 6.6 and 12.0 kg/ha, respectively. Pesticides are still preferred by small farmers because they are cost-effective, easily available, and display a wide spectrum of bioactivity. Out of the total consumption of pesticides, 80% are in the form of insecticides, 15% are herbicides, 1.46% are fungicides and less than 3% are others.

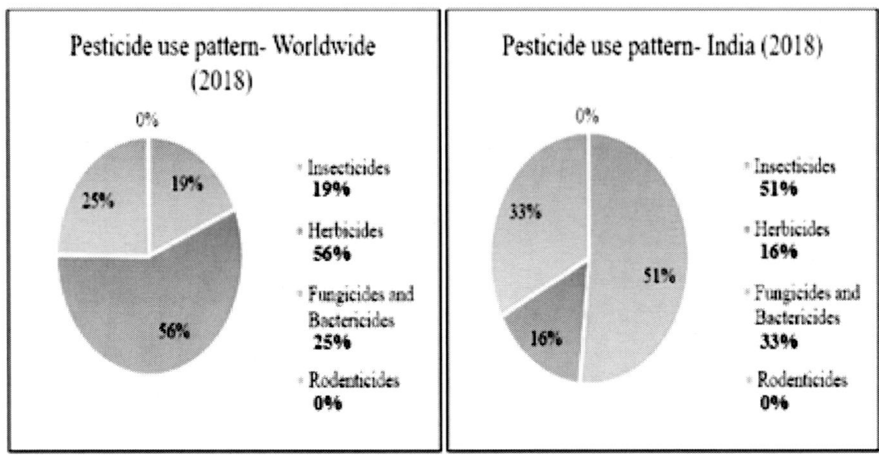

Source: http://www.fao.org/faostat/en/#data) (FAO, 2018).

Figure 1b. Pesticide usage in World and India 2018.

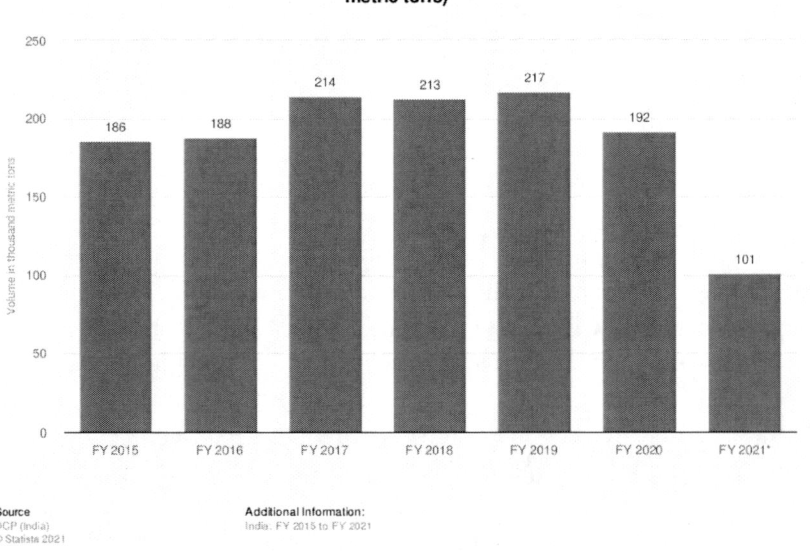

Figure 1c. India: Pesticides Production volume 2021/Statista.

In comparison, the worldwide consumption of herbicides is 47.5%, insecticides 29.5%, fungicides 17.5% and others account for 5.5% (Gupta, 2004). The consumption of herbicides in India is probably low because weed control is mainly done by hand weeding. Furthermore, the Food and Agricultural Organization of the United Nations (FAO) has estimated that more than 400,000 tonnes of obsolete pesticides are stocked worldwide (FAO, 2000; Vlyssides et al. 2004). Among the various pesticides used in India, 40% of all the pesticides used belong to an organochlorine class of chemical pesticides. The Figures 1b, 1c and 1d, represents the Indian scenario of pesticide usage.

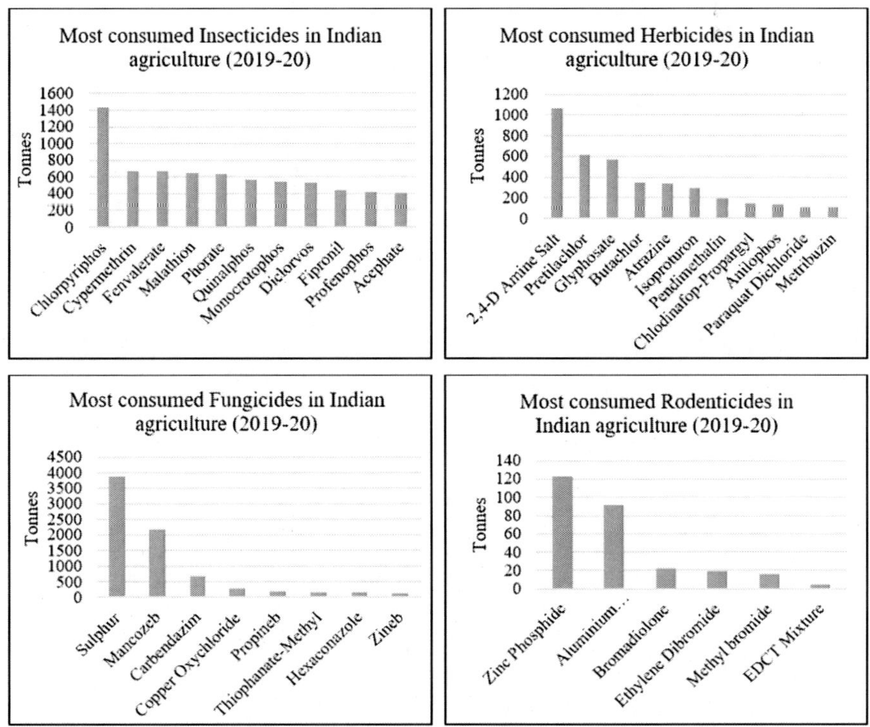

Figure 1d. Most consumed Insecticides, herbicides, fungicides, and rodenticides during 2019-20 in India (Nayak, P., & Solanki, H., 2021).

The other major category is organophosphate pesticides, Monocrotophos, phorate, phosphamidon, methylparathion, and dimethoate are some of the highly hazardous pesticides that are continually and indiscriminately used in India. The higher consumption of insecticide is partially due to the warm humid and tropical climate which provides a favorable breeding environment for insects coupled with a shorter life cycle and higher hatching rate.

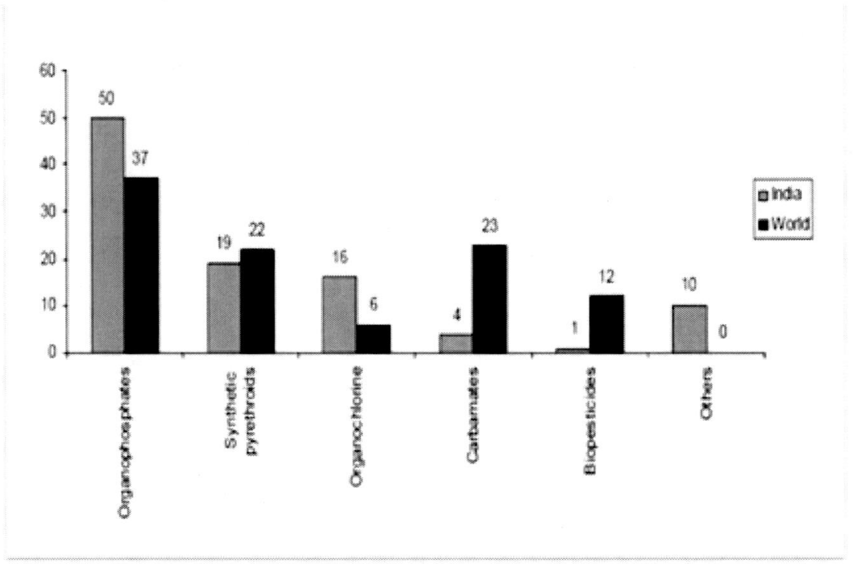

Centre for Science and Environment (2003)

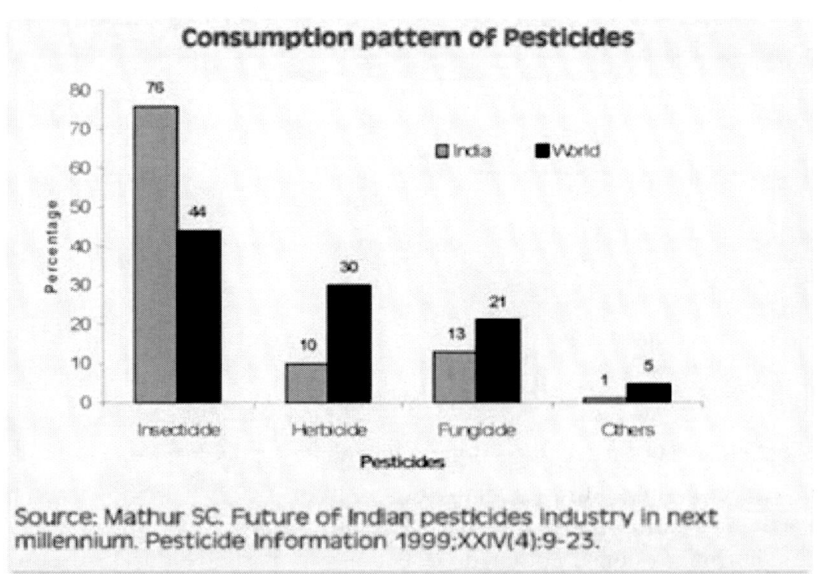

Figure 1e. Pesticide Consumption in India and the World. (Centre for Science and Environment, 2003).

Among the pesticide usage patterns, the insecticide usage was higher and the organophosphate class of insecticides was widely used in India. Moreover, the organophosphate residues were found even in soft drinks and were reported by the Centre for Science and Environment in New Delhi, (CSE, 2003) (Figure 1e). The widespread use of pesticides for agricultural and non-agricultural purposes has resulted in the presence of their residues in various environmental matrices. Pesticide contamination of surface water has been well documented worldwide and constitutes a major issue that gives rise to concerns at local, regional, national, and global scales (Figures 1a, 1b, 1c, 1d and 1e).

Chapter 4

Organophosphate Pesticides: An Overview

Recent studies have shown that home environments are commonly contaminated with pesticides especially organophosphates which are the most commonly used pesticides (Eskenazi et al. 1999). In addition to large one-time exposures, the pesticide residues from applications around the home, as well as residues on food and in drinking water can lead to long-term, low-level exposure to organophosphates. In health, agriculture, and government, the word "organophosphates" refers to a group of insecticides or nerve agents acting on the enzyme acetylcholinesterase (the pesticide group Carbamates also act on this enzyme but through a different mechanism). The term is used often to describe virtually any organic phosphorus (V) containing compound, especially when dealing with neurotoxins. Many of the so-called organophosphates contain C-P bonds. For instance, sarin is O-isopropyl methylphosphono fluoridate, which is formally derived from HP (O) $(OH)_2$, not phosphoric acid. Also, many compounds which are derivatives of phosphinic acid are used as organic phosphorus-containing neurotoxins.

Organophosphate pesticides (as well as Sarin and VX nerve gas) irreversibly inactivate acetylcholinesterase, which is essential to nerve function in insects, humans, and many other animals. Organophosphate pesticides affect this enzyme in varied ways, and thus in their potential for poisoning. For instance, parathion, one of the first organophosphate pesticides commercialized, is many times more potent than Malathion, an insecticide used in combating the Mediterranean fruit fly (Med-fly) and West Nile Virus-transmitting mosquitoes. Recent analyses of groundwater have indicated that the general population is not adequately protected from pesticide contamination (Walker and Keasling, 2002) and that pesticides are found in groundwater sooner than generally can be predicted with traditional management models (Pivetz et al. 1996). Among known pesticide groups, organophosphorus compounds have grasped attention due to their hazardous nature and their applications as pesticides, and chemical weapons this group of pesticides has a broad spectrum of applications for crops and animal protection as well as for public health. Such pesticides are essential for most agriculture-developing programs. However, the extensive use and/or misuse of this group of pesticides in different ways has resulted in many undesirable effects and has caused substantial environmental pollution problems. The World Health Organization reported that exposure to such pesticides results in roughly 3 million cases of severe poisoning and 220,000 deaths annually worldwide. Also, more than 0.7 million cases of poisoning worldwide could be traced to occupational and incidental exposure to Organophosphorus (Gill and Ballesteros, 2000).

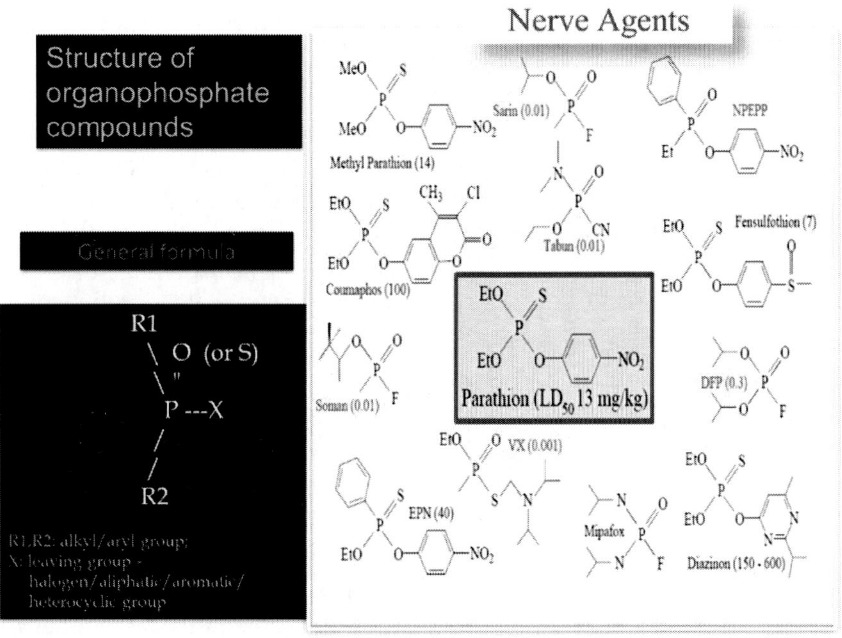

Figure 2a. Figure Structure of Organophosphate pesticides.

India is an agriculture-based country, about 30% of agricultural produce is lost due to pests. Hence agrochemicals like insecticides, fungicides, pesticides, and herbicides have increasing demand for the protection of crops in one way or the other (Anonymous 2000). Thus, large numbers of pesticides are emerging day by day. Furthermore, the Food and Agricultural Organization of the United Nations (FAO) has estimated that more than 400,000 tonnes of obsolete pesticides are stocked worldwide (FAO, 2000 Vlyssides et al. 2004). The disposal of EDCs such as pesticides, fungicides, and insecticides can cause serious environmental problems due to the chemical toxicity of the active ingredients in the formulations and sometimes, of their decomposition products. Even though many insecticides degrade rapidly in the soil, they can be potentially hazardous as a consequence of accidental spills, runoff from applications in agricultural areas, and discharge from pesticide containers and waste disposal systems (Rani and Lalithakumari, 1994). Hertel (1993), Shimazu et al. (2001) and Zhongli et al. (2001) mention that organophosphate (OP) compounds represent the largest group of chemical insecticides used in plant protection throughout the world. In the United States, OPs represent about half of the total insecticides used, with annual applications of over 75 million pounds (Bravo et al. 2002). Because of the rapid growth in industrial and agricultural chemical usage, vast quantities of soil and groundwater have been contaminated with hazardous compounds (Chauhan et al. 2000). Recent analyses of groundwater have indicated that the general population is not adequately protected from pesticide contamination (Walker and Keasling, 2002) and that pesticides are found in groundwater sooner than generally can be predicted with traditional management models (Pivetz et al. 1996). Among known pesticide groups, organophosphorus (OP) compounds have grasped attention due to their hazardous nature and their applications as pesticides and chemical weapons. This group of pesticides has a broad spectrum of applications for crops and animal protection as well as for public health. Such pesticides are essential for most agriculture-developing programs. However, the extensive use and/or misuse of this group of pesticides in different ways has resulted in many undesirable effects and has caused substantial environmental pollution problems. The World Health Organization reported that exposure to such pesticides results in roughly 3 million cases of severe poisoning and 220,000 deaths annually worldwide. Also, more than 0.7 million cases of poisoning worldwide could be traced to occupational and incidental exposure to OPs (Gill and Ballesteros, 2000).

Organophosphate Pesticides are all called "non-persistent pesticides" because they break down fairly rapidly in the environment (within days or weeks), reducing their potential to accumulate in the tissues of plants, animals, or humans. This class is the most widely used type of pesticide, with common uses in agriculture, in the home, and in the garden. Organophosphate compounds make up the largest class of insecticides currently used in industrial countries (Yang et al. 1995). These pesticides are used worldwide to control insects in a variety of diverse applications, plasticizers, air-fuel ingredients chemical warfare agents, and drugs for the treatment of diseases such as glaucoma, or parasitic infections (Chatonnet et al. 1999). Organophosphate pesticides such as monocrotophos, chloropyriphos, parathion, methyl parathion, diazinon, and coumaphos are perhaps the most extensively used insecticides in many agricultural practices (Singh et al. 1999). They are the most widely used insecticides, accounting for an estimated 34% of worldwide insecticide sales. Contamination of soil from pesticides as a result of their bulk handling at the farmyard or following application in the field or accidental release may lead occasionally to contamination of surface and groundwater. Several reports suggest that a wide range of water and terrestrial ecosystems may be contaminated with organophosphorus compounds. These compounds possess high mammalian toxicity and it is therefore essential to remove them from the environments. In addition, about 200,000 metric tons of nerve (chemical warfare) agents have to be destroyed worldwide under the Chemical Weapons Convention (1993). Organophosphate includes paraoxon, parathion, chlorpyrifos disulfolon, redene, carbophenothion, and dimeton. The neurotoxicological properties of this class of compounds are namely due to its ability to suppress acetylcholinesterase and as a result, prevent acetylcholine esterase from breaking down acetylcholine at the synaptic junction. The Acetylcholine (Ach) and Acetylcholine esterase (AChE) activity is depicted in Figure 2. Thus they irreversibly inhibit acetylcholinesterase and these compounds, have also been associated with pathology and chromosomal damage associated with bladder cancer (Webster et al. 2002).

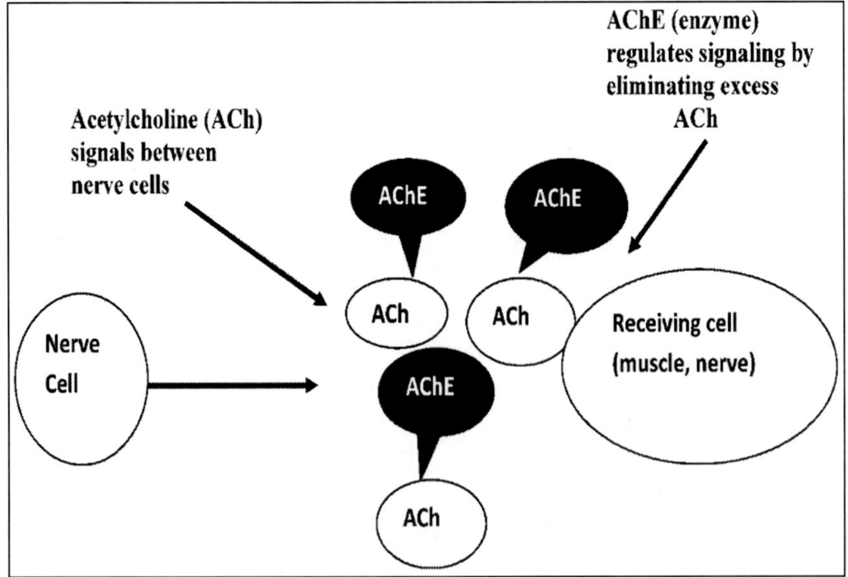

Figure 2b. Acetylcholine (Ach) and Acetylcholine esterase (AChE) activity.

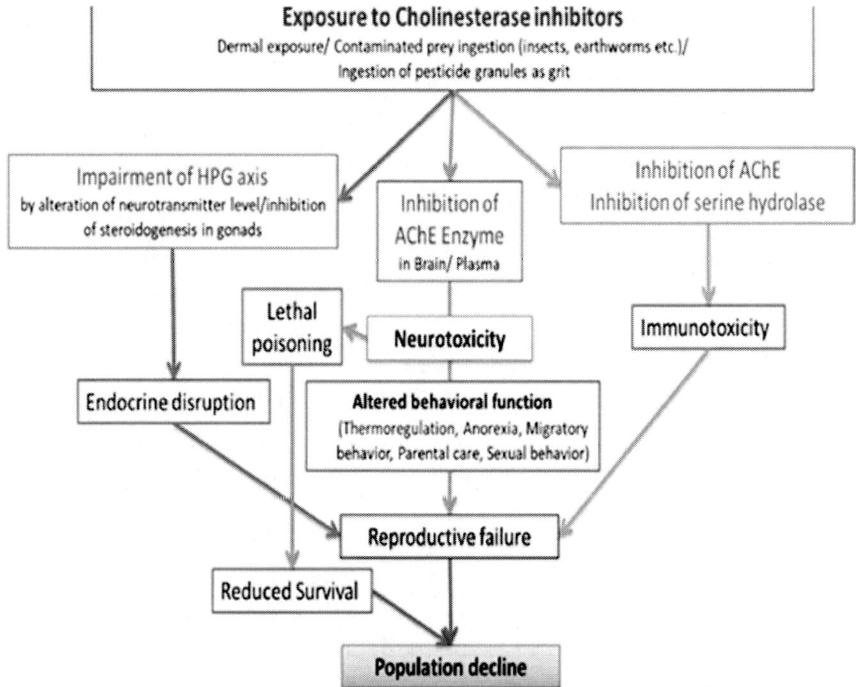

Figure 2c. Exposure to Choline esterase inhibitors and Endocrine disruption.

Many nitroaromatic compounds are widely used in the manufacture of explosives, pesticides, and dyes. For example, 4-nitrophenol and 2,4-dinitrophenol are commodity chemicals commonly used in the manufacture of pesticides and pharmaceuticals. Additionally, 4-nitrophenol is released into the environment during the hydrolysis of several organophosphorous pesticides such as parathion (Pitter and Chudoba, 1990). Organophosphorus (OP) compounds with carbon-phosphorus bonds are widespread among man-made chemical substances- xenobiotics, which uncontrollably enter the environment and become a toxic factor (Kononova and Nesmeyanova, 2003). Organophosphates have been extensively applied as alternatives to organochlorine compounds which possess long-term persistence and high toxicity. But, organophosphorus compounds rapidly undergo degradation by soil microorganisms, so they do not persist in the environment. However, repeated applications of degradable organophosphates occasionally cause a significant reduction of their pesticidal effect. This phenomenon results from microbial adaptation to pesticide degradation. OP pesticides are synthetic esters, amides, or thiol derivatives of phosphoric, phosphonic, phosphorothioic, or phosphonothioic acids. There are over 100 OP compounds currently on the market, representing a variety of chemical, physical, and biological properties (Sherine et al. 2010).

The organophosphate pesticides include butamifos, chlorfenvinphos, chlorpyrifos, chlorpyrifosmethyl, cyanofenphos, cyanophos, diazinon, dichlorvos, dimethoate, dioxabenzofos, edifenphos, EPN, ethion, fenitrothion, fenthion, glyphosate, isofenphos, isoxathion, malathion, methidathion, methyl-parathion, mevinphos, parathion, phenthoate (fenthoate), phorate, phosmet, pirimiphos- methyl, tetrachlorvinphos, and tolclofos-methyl., etc. The neurotoxicological properties of this class of compounds are namely due to its ability to suppress acetylcholinesterase and as a result, prevent acetylcholine esterase from breaking down acetylcholine at the synaptic junction. Thus they irreversibly inhibit acetylcholinesterase and these compounds, have also been associated with pathology and chromosomal damage associated with bladder cancer.

Many nitroaromatic compounds are widely used in the manufacture of explosives, pesticides, and dyes. For example, 4-nitrophenol and 2,4-dinitrophenol are commodity chemicals commonly used in the manufacture of pesticides and pharmaceuticals. Additionally, 4-nitrophenol is released into the environment during the hydrolysis of several organophosphorus pesticides such as parathion (Pitter and Chudoba, 1990). Organophosphate pesticides degrade rapidly by hydrolysis on exposure to sunlight, air, and soil, although small amounts can be detected in food and drinking water. The ability to degrade made organophosphate an attractive alternative to persistent organochlorine pesticides. Although organophosphates degrade fasten the organochlorines, they have greater acute toxicity, posing risks to people who may be exposed to large amounts. The exposure to Cholinesterase inhibitors and Endocrine disruption is shown in Figure 2b.

4.1. Major Sources of Pesticide Water Pollution

Wastewater from agricultural industries and pesticide formulating or manufacturing plants were reported to have pesticide contamination levels as high as 500 mgL^{-1} (Serge et al. 2000). There are four major sources of pesticide water pollution (Chiron et al. 1997).

- Pesticide treatment as a consequence of agricultural practices (a concentration range of few ppb),
- Rinse water from containers and spray equipment (10-100 ppm),
- Wastewater from agricultural industries (10-100 ppm),
- Wastewater from formulating or manufacturing pesticide plants (1-1000 ppm).

Organophosphorus compounds have overtaken organochlorine compounds as the most used insecticides in the recent decade. Among the organophosphate insecticides, methylparathion was found to be a widely used insecticide in India. Its widespread use has caused environmental concern due to its frequent leakage into surface and ground waters. A maximum of 0.1-1.0 mgL^{-1} of methylparathion is allowed for the crops (fruits, vegetables, nuts, and grains) to be used for human consumption. According to the Environmental Protection Agency, the following levels of methylparathion in drinking water are not expected to cause effects that are harmful to health: 0.3mgL^{-1} for 1 or 10 days of exposure to children, 0.03 mgL^{-1} on longer-term exposure to children and 0.002 mgL^{-1} on lifetime exposure to adults. Although methylparathion is a less persistent insecticide, the hydrolysis product p-nitrophenol, is more toxic than its parent compound (Barik, 1984) and it has a great environmental concern since p-nitrophenol is considered to be a priority pollutant (Keith and Telliard, 1979). Therefore, there is a need for economically dependable methods of Organophosphorus detoxification from the environment.

Chapter 5

Chemistry of Organophosphates

Organophosphates are widely employed both in natural and synthetic applications because of the ease with which organic groups can be linked together. Being a triprotic acid, phosphoric acid can form *triesters* whereas carboxylic acids only form monoesters. Esterification entails the attachment of organic groups to phosphorus through oxygen linkers. The precursors to such esters are *alcohols.* Encompassing many thousands of natural and synthetic compounds, alcohols are diverse and widespread.

$$OP\,(OH)_3 + ROH \rightarrow OP\,(OH)_2(OR) + H_2O \quad (1.1)$$

$$OP\,(OH)_2(OR) + R'OH \rightarrow OP\,(OH)\,(OR)\,(OR') + H_2O \quad (1.2)$$

$$OP\,(OH)\,(OR)\,(OR') + R''OH \rightarrow OP\,(OR)\,(OR')\,(OR'') + H_2O \quad (1.3)$$

The phosphate esters bearing OH groups are *acidic* and partially *deprotonated in* aqueous solution (Eqns. 1.1 - 1.3). For example, DNA and RNA are polymers of the type $[PO_2\,(OR)\,(OR')^-]_n$. Polyphosphates also form esters. An important example of an ester of polyphosphate is *ATP,* which is the monoester of triphosphoric acid ($H_5P_3O_{10}$). Alcohols can be detached from phosphate esters by *hydrolysis,* which is the reverse of the above reactions. For this reason, phosphate esters are common carriers of organic groups in *biosynthesis.*

5.1. Structural Features of Organophosphate Neurotoxins: Chemical Structure

The General formula for organophosphorus compounds is shown in Figures 3 and 4.

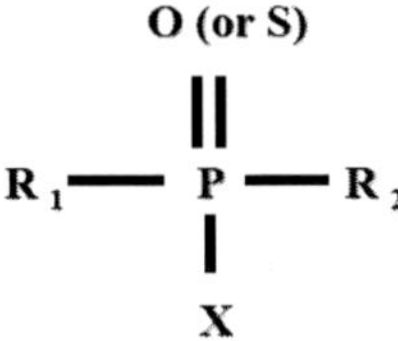

Figure 3. General chemical structure of organophosphorus compounds.

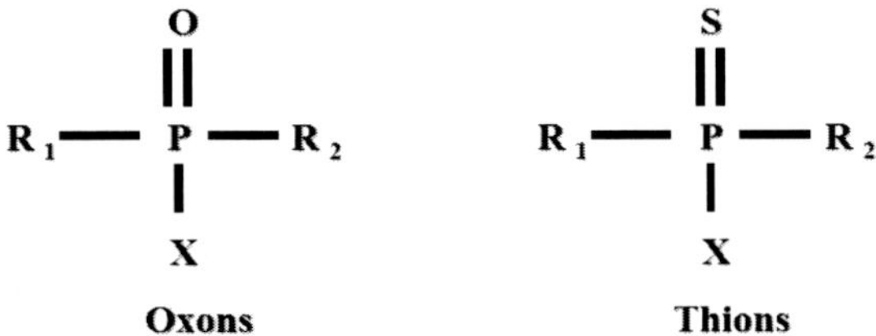

Figure 4. General chemical structure of oxon and thion.

Structurally, both oxons and thions show variety in the single-bonded R_1, R_2 and X groups attached to the central pentavalent phosphorus atom. However, R_1 and R_2 generally tend to be alkoxy, aryloxy, and thioalkoxy groups, while X is a labile leaving group. Tetraethyl pyrophosphate (TEPP), the most potent oxon pesticide, has a unique pyrophosphate structural motif, a biochemically important high-energy phosphate bond. The most potent thion pesticide, parathion, has a *p*-nitrophenoxy (X) substituent, which is a very good leaving group in nucleophilic substitution reactions involving the central phosphorus atom (Sherine et al. 2010).

5.2. Effective Organophosphate Neurotoxins and Their Structural Features

A terminal oxygen is connected to phosphorus by a double bond, i.e., a phosphoryl group and two lipophilic groups bonded to the phosphorus. A leaving group bonded to the phosphorus, often a halide and terminal oxygen vs. terminal sulfur are the features found in the structure. Thiophosphoryl compounds, those bearing the P=S functionality, are much less toxic than related phosphoryl derivatives, which include sarin, VX, and tetraethyl pyrophosphate. Thiophosphoryl compound is not an active inhibitor of acetylcholinesterase in either mammals or insects, in mammals the animal's metabolism tends to remove lipophilic side groups from the phosphorus atom while an insect tends to oxidize the compound so removing the terminal sulfur and replacing it with a terminal oxygen which causes the compound to be more able to act as an acetylcholinesterase inhibitor. Within these requirements, a large number of different lipophilic and leaving groups have been used. The variation of these groups is one means of fine-tuning the toxicity of the compound. A good example of this chemistry is the *P*-thiocyanate compounds which use an aryl (or alkyl) group and an alkyl amino group as the lipophilic groups. The thiocyanate is the leaving group. It was claimed in a German patent that the reaction of 1, 3, 2, 4-dithiadiphosphetane 2, 4-disulfides with dialkyl cyanamides formed plant protection agents that contained six-membered (P-N=C-N=C-S-) rings. It has been proven in recent times by the reaction of di-ferrocenyl 1,3,2,4-dithiadiphosphetane 2,4-disulfide (and Lawesson's reagent) with dimethyl cyanamide that a mixture of several different phosphorus-containing compounds is formed. Depending on the concentration of the dimethyl cyanamide in the reaction mixture, either a different six-membered ring compound (P-N=C-S-C=N-) or a nonheterocyclic compound (FcP(S) (NR_2) (NCS)) is formed as the major product; the other compound is formed as a minor product. In addition, small traces of other compounds are also formed in the reaction. It is unlikely that the ring compound (P-N=C-S-C=N-) (or its isomer) would act as a plant protection agent, but (FcP(S) (NR_2)(NCS)) compounds can act as nerve poisons in insects (www.medlibraryorg).

Chapter 6

Methylparathion

Methylparathion is one of the most widely used organophosphate insecticides. It is used for the control of aphids, caterpillars, elicoverpa spp, moths, butterflies, mites, weevils, jassids, and budworms in the cultivation of fruits, vegetables, cotton, and tobacco. Its widespread use has caused environmental concern due to its frequent leakage into surface and ground waters. It is acutely toxic to mammals, for example, the oral LD_{50} for rats is in the range of 18–50 mg kg^{-1} (Extoxnet, Hayes and Laws, 1990; Occupational Health Services, Inc. 1991 and Meister, 1992). It acts by inhibiting the enzyme acetylcholinesterase in nerve tissue. Because of their high biological activity, and in some cases persistence in the environment, the use of pesticides may cause undesired effects on human health and to environment. Organophosphorus compounds have overtaken organochlorine compounds as the most used insecticides in the recent decade. Methyl parathion is an endocrine disruptor. The US EPA (2003) reported possible endocrine disruption in mammals. The ATSDR (2001) also reported indications of weak oestrogenic activity. It has shown oestrogenic potential similar to 17beta-estradiol (the primary natural estrogen) in trout cells (Petit et al. 1997), and it induces the activity of aromatase, an enzyme that converts androgens to estrogen thus increasing breast cancer risk (Laville et al. 2006). Exposure to a mixture of methylparathion and chlorpyrifos at 1/30th the LD_{50} affected endocrine hormone levels in rats, increasing oestradiol in both males and females (Liu et al. 2006). It caused increased testosterone and decreased luteinizing hormone in the testes (Narayana et al. 2006a).

6.1. Chemical Identity

Chemical name:	O, O-Dimethyl O-(4-nitrophenyl) phosphorothioate
Common name:	Parathion methyl
Manufacturer's code	
CAS Registry number:	298-00-0

Molecular formula: $C_8H_{10}NO_5PS$
Molecular weight: 263.206
Molecular Structure: as shown in Figure 5

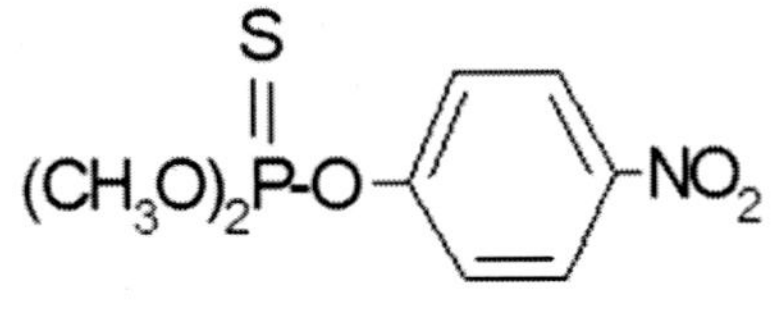

Figure 5. Molecular structure of methylparathion.

6.2. Physico-Chemical Properties

The following are the physicochemical properties of methylparathion.

Appearance:	White crystals or a brown amber liquid
Odour:	Rotten eggs or pungent garlic
Melting Point:	35-36°C
Specific Gravity/Density:	1.358 @ 20°C
Vapour Pressure:	1.3×10^{-5} mbar @ 20°C
Solubility in Water:	70.3 mgL^{-1} @ 25°C
Octanol-Water Coefficient:	Log Pow = 2.8
Henry's Constant:	8.45×10^{-8} atm m^3/gmol
Dissociation Constant:	Not relevant. There are no dissociable hydrogens.

Methylparathion (*O*-dimethyl *O*-(4-nitro-phenyl) phosphorothioate) has a molecular formula of $C_8H_{10}NO_5PS$ with a molecular mass of 263.23. Pure methylparathion is a white crystalline solid or powder; technical (80%) grade is a light to dark tan liquid. Pure methylparathion solubility in water is 55-60 mgL^{-1} at 25°C; it is readily soluble in most organic solvents, soluble in ethanol, chloroform, and aliphatic solvents, and slightly soluble in light petroleum. The odor of methylparathion is like rotten eggs or garlic. Hydrolysis of methyl parathion occurs in different conditions but it is more rapid under alkali conditions (Kuo and Perera, 2000). The hydrolysis and biodegradation were studied in four types of water (ultra-pure, pH 6.1; river water, pH 7.3; filtered river water, pH 7.3; and seawater, pH 8.1) maintained at 6ºC and 22ºC in the dark by Zheng and Liu (2002) and Serdar and Gibson (1985). They reported the half-lives of methylparathion at 6ºC in the four types of water were determined to be 237, 95, 173, and 233 days, and at 22ºC, the half-lives were 46, 23, 18, and 30 days respectively. Technical methylparathion is available as a solution containing 80% active ingredient, 16.7% xylene, and 3.3% inert ingredients (Hertel, 1993). It is one of the highly active thiophosphorus ester insecticides developed in the 1940s by the German pesticide company Bayer. It is produced worldwide by many companies. The production of methylparathion in 1966 was 31,700 tonnes, from which 14,800 tonnes were produced in the United States (Hertel, 1993). It is registered in at least 38 countries and widely used throughout the world.

6.3. Water Solubility

Determination of water solubility in pure water was carried out in compliance with the US Environmental Protection Agency. Several deviations from the protocol were noted. These deviations did not invalidate the test results. Radio-labeled material and non-labeled analytical grade (99.5%) material was mixed with deionized water and maintained at 25°C with constant agitation. The test solution was sampled every 25 hours for 4 days and solutions were analyzed by liquid scintillation. Equilibrium was reached in the first 24-hour period. The water solubility was determined as 70.3 ± 2.73 mgL^{-1} at 25°C.

Methylparathion is used to control chewing and sucking insects such as aphids, boll weevils, and mites in a wide variety of crops, including cereals, fruits, vines, vegetables, ornamentals, cotton, and field crops (Pesticide News, 1995; Garcia et al. 2003). It is a broad-spectrum non-systemic pesticide that kills pests by stomach poisoning, is an acaricide, and has some fumigant action (Garcia et al. 2003; Hertel, 1993). Methylparathion was originally registered in the United States in 1954. However, its use has been restricted in this country since 1978. Methylparathion is used throughout the southern and Midwestern parts of the United States. As a restricted-use pesticide, the application of methylparathion requires appropriately trained certified pesticide applicators. Recently, the use of methylparathion has voluntarily been canceled in many of the most significant food crops by the United States Environmental Protection Agency (EPA). In the United States, annual agricultural application peaked at 12,503 metric tons in 1971 and was 3,471 metric tons in 1982 (Garcia et al. 2003). Contamination of fields, crops, water, and air occurs through off-target spraying as a consequence of agricultural and forestry activities. For this reason, exposure may occur through air, water, and food-borne residues of methylparathion (Hertel, 1993). Methylparathion is not very persistent in the environment; it is not bioconcentrated and is not transferred through food chains. It is efficiently degraded by several microorganisms and other forms of wildlife. This insecticide is likely to cause damage to ecosystems only in instances of heavy over-exposure resulting from accidental spills (Hertel, 1993).

Degradation of methylparathion appears to be faster in the presence of sediment and freshwater rather than in salt water. The rate of degradation depends on the presence and acclimation of microbial populations in the body of water (Extoxnet, 1996). It has a half-life in water environments of 175 days, and 10 days to two months in soils. Temperature and exposure to sunlight increase the rate of degradation. When large concentrations of methylparathion reach the soil, for instance in an accidental spill, degradation will occur only after many years (Pesticide News, 1995). It does not volatilize significantly at ambient conditions (Extoxnet, 1996). Oxidative degradation of methylparathion to the less stable methyl paraoxon may occur by ultraviolet radiation or sunlight. It also undergoes chemical hydrolysis in aquatic environments; however, this degradation reaction is not a significant contribution to the disappearance of methylparathion. In towns in the center of agricultural areas in the United States, methylparathion concentrations in natural waters ranged up to 0.46 μg/L, with the highest levels found during the summer (Hertel, 1993).

Chapter 7

Health Effects of Organophosphate Compounds

The primary biochemical effect associated with high-level exposure to organophosphate insecticides is the inhibition of acetylcholinesterase (AChE), resulting in acetylcholine accumulation (Hertel, 1993 Shimazu et al. 2001). The normal function of AChE is to terminate neurotransmission due to acetylcholine; it is liberated at cholinergic nerve endings in response to nervous stimulation. Loss of AChE activity may lead to a variety of effects resulting from excessive nervous stimulation, culminating in respiratory failure and death. Hypotension, bradycardia, bronchoconstriction, and bronchial fluid accumulation, symptoms that result from the inability of respiratory muscles to work, are observed following lethal amounts of methylparathion. Cyanosis and central respiratory depression can be observed as well. In less severe cases of intoxication, bradycardia, muscle rigidity, muscle hypotonia, bronchial spasm, and constriction might occur (Hertel, 1993). Clinical symptoms of poisoning with methylparathion include pallor, sweating, dizziness, vomiting, diarrhea, abdominal cramps, headache, blurred vision, convulsions, dilation of the pupils, tears, salivation, cardiac arrest, and in extreme cases death. Methylparathion is a highly toxic organophosphate ester insecticide (Hertel, 1993). It is easily absorbed via all routes of exposure (oral, dermal, inhalation) and is rapidly distributed to the tissues of the body. Conversion of methylparathion to methyl paraoxon occurs within minutes of administration. Methyl paraoxon is 10 times more toxic than methylparathion. Metabolism and detoxification occur through the liver. The mechanisms of detoxification of methylparathion or methyl paraoxon are mainly through oxidation, hydrolysis, and demethylation or dearylation with reduced glutathione (Hertel, 1993). Effects such as impaired memory and concentration, disorientation, severe depression, irritability, confusion, headache, speech difficulties, delayed reaction times, nightmares, sleepwalking, drowsiness, and insomnia have been reported in workers repeatedly exposed to methylparathion. Studies suggest that there is a decrease in blood cholinesterase activity without clinical manifestations following repeated, long-term exposures (Hertel, 1993). Some organophosphate compounds also may cause delayed neurotoxicity called OP-induced delayed neuropathy (OPIDN) (Cho, 2001). Methylparathion has also been reported to have DNA-alkylating properties. The results of mutagenicity tests have been both positive and negative. The results of most of the *in vitro* mutagenicity studies with both bacterial and mammalian cells were positive; whereas, the *in vivo* studies produced equivocal results (Hertel, 1993).

Chapter 8

Organophosphorus Pesticide Compounds and Their Toxicity

The pesticides are described as chemicals that kill or slow down the growth of undesirable organisms. These pesticides include herbicides, insecticides, fungicides, and nematocides. Nowadays, it is believed that the application of synthetic pesticides is one of the most effective methods for controlling insects that affect crop growth (Celik et al. 1995). Organophosphorus pesticides constitute the most widely used insecticides available today. This class of compounds has achieved enormous commercial success as a key component in the arsenal of agrochemicals and is currently an integral element of modern agriculture across the globe. According to the EPA, about 70% of the insecticides in current use in the US are organophosphorus pesticides (U.S. EPA, 2010). They were developed to replace organohalide pesticides in the late 1950s because organophosphorus pesticides are relatively easier to degrade via microbial or environmental processes. Unlike organohalide pesticides, organophosphorus pesticides do not bioaccumulate due to their rapid breakdown in the environment and they are thus preferred over organohalides for insecticide and/or pesticide use (Sherine et al. 2010). Although organophosphorus compounds are considered safer than organohalides, they are still highly neurotoxic to humans, and in some cases, their degradation products have the potential to be more toxic with chronic exposure. The organophosphorus compounds (OP) are efficiently absorbed by inhalation, ingestion, and skin penetration. They are strong inhibitors of cholinesterase enzymes that function as neurotransmitters, including acetylcholinesterase, butylcholinesterase, and pseudocholinesterase. These enzymes are inhibited by binding to the organophosphorus compound. Upon binding, the organophosphorus compound undergoes hydrolysis leading to a stable phosphorylated and a largely unreacted enzyme. This inhibition results in the accumulation of acetylcholine at the neuron/neuron and neuron/muscle junctions or synapses (Sherine et al. 2010).

Each year, organophosphorus pesticides poison thousands of humans across the world. In fact, in 1994, an estimated 74,000 children were involved in common household pesticide-related poisoning or exposures in the United States (U.S. EPA, 2010). In a more recent study, it was found that children exposed to organophosphorus pesticides were more likely to be diagnosed with attention deficit hyperactivity disorder (ADHD) (Bouchard et al. 2010). The exposure has been attributed to the frequent use of organophosphorus pesticides in agricultural lands and their presence as residues in fruits, vegetables, livestock, poultry products, and municipal aquifers (Liu and Lin, 1995). For example, typical pesticide concentrations that flow into an aqueous waste range from 10,000 ppm to 1ppm (Survey, 2010, http://ga.water.usgs.gov/publications/ofr00-187.pdf and Gilliom et al. 1999).

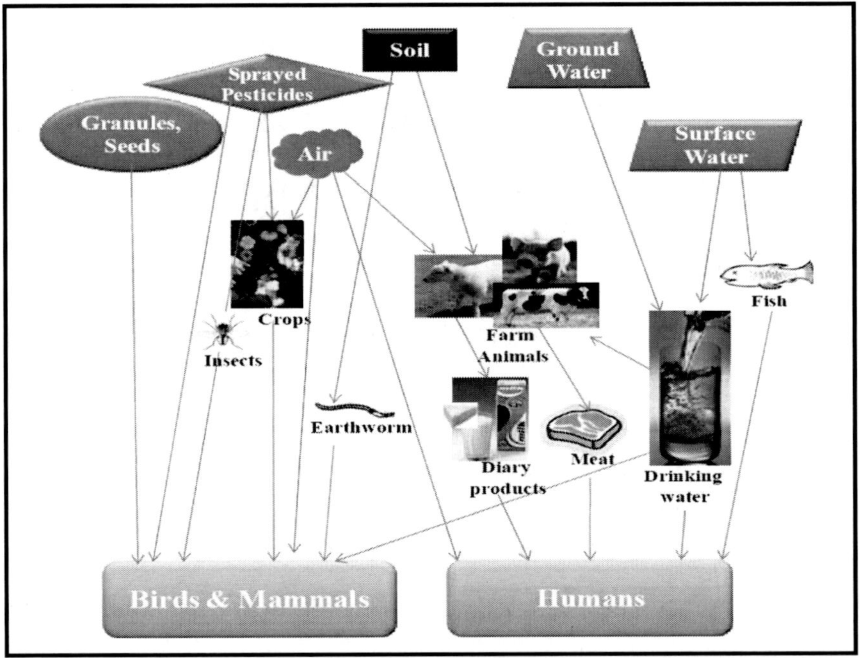

Figure 6. Schematic representation of the possible routes of environmental exposure of organophosphorus pesticides to humans and wildlife (Adopted from Reference Vermeire et al. 2003).

These pesticides are influenced by several biological, chemical, and physical processes once they enter the environment. Figure 6, shows the possible routes of environmental exposure to organophosphorus pesticides to humans and wildlife (Vermeire et al. 2003). While many OP pesticides can degrade via microbial or environmental processes, some of these pesticides are consumed by organisms, or they could leach into groundwater. Once a pesticide enters ground water it can remain there for considerable periods. In groundwater, there is little sunlight exposure, which slows down the degradation of OP pesticides and increases their potential risks to the environment and human health.

The first indication of insecticidal activity among OP compounds was found in 1930, however, the first compound of this type, hexamethyl tetraphosphate (HETP) [an impure mixture containing tetraethyl pyrophosphate (TEPP) as the active ingredient] was not used as an agricultural insecticide until 1942 (Walker., 1972). Since the first introduction of HETP, the number of OP pesticides has risen to hundreds, and the common ones are shown in Table 1 and Table 2, along with their toxicity information. As indicated by their LD_{50} values in Table 2, there are some significant bioactivity differences between oxons and thions. Generally, compared to the pesticides, the oxons are more potent with lower LD_{50}'s. One of the most potent oxon pesticides is TEPP with an oral LD_{50} of 0.5 mg/kg, while one of the most active thion pesticides is parathion with an oral LD_{50} of 1 mg/kg (Sherine et al. 2010). In general, Table 2, indicates that the oral toxicity for an individual OP is much greater than their dermal toxicity.

Table 1b. LD Dosage for Methylparathion and Chlorpyrifos in Rats

Interactive toxicity of chlorpyrifos and methyl parathion

Dosages ($\times LD_1$)[a]	First exposure	4-Day lethality
0.5	Chlorpyrifos	1/8
	Methyl parathion	0/8
1.0	Chlorpyrifos	8/8
	Methyl parathion	0/8

[a] The respective LD_1 dosages of chlorpyrifos and methyl parathion were 80 and 4 mg/kg, p.o., respectively. Rats ($n = 8$ per treatment group) were given one OP insecticide (First exposure) by gavage in peanut oil (1 ml/kg) 4 h prior to the second OP and cumulative lethality was recorded for 4 days (from Karanth et al., 2004).

Table 2. Common organophosphate (OP) pesticides are classified by oxon/thion structure and oral LD50 toxicities (Walker, 1972; Gilliom et al. 1999; Vermeire et al. 2003)

No	OP Name	Structure (Thions: 1–17; Oxons: 18–29)	LD_{50}, mg/kg *		WHO Acute Hazard §	IARC Carcinogens‡	U.S. EPA Carcinogens†
			Oral	Dermal			
1	Parathion		1	21	Ia	Unclassifiable	Possible
2	Fonofos		8–17	147	Ia	N/A	Unlikely
3	Azinphos-methyl		11–13	220	Ib	N/A	Not Likely
4	Coumaphos		16–41	1,000	Ib	N/A	Not Likely
5	Methidathion		25–48	1,546	Ib	N/A	Possible
6	Leptophos		45–53	>800	N/A	N/A	N/A
7	Propetamphos		75–82	2,300	Ib	N/A	Not Likely
8	Carbophenothion		98–120	190–215	N/A	N/A	N/A
9	Phosmet		113–160	>1,500	II	N/A	Suggestive
10	Chlorpyrifos		135–165	2,000	II	N/A	Unlikely

11	Fenthion		214–245	330	II	N/A	unlikely
12	Fenitrothion		250	>3,000	II	N/A	unlikely
13	Dichlofenthion		270	6,000	N/A	N/A	N/A
14	Dicapthon		330–400	790–1,250	N/A	N/A	N/A
15	Diazinon		300–850	2,150	II	N/A	Not Likely
16	Ronnel		1,250–2,630	2,000	N/A	N/A	N/A
17	Malathion		5,400–5,700	>2,000	III	Unclassifiable	Suggestive
18	Tetraethyl pyrophosphate (TEPP)		0.5	2.4	N/A	N/A	N/A
19	Mevinphos		3.7–6.1	4.2–2.7	Ia	N/A	N/A
20	Schradan		10	15	N/A	N/A	N/A

Table 2. (Continued)

21	Monocrotophos	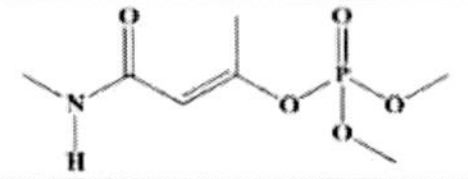	18–20	112–126	Ib	N/A	N/A
22	Phosphamidon	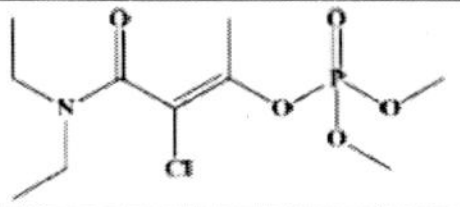	24	107–143	Ia	N/A	Possible
23	Oxydemeton methyl	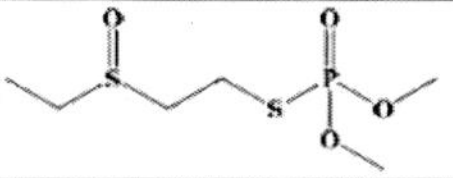	47–52	158–173	Ib	N/A	Not Likely
24	Ethoprophos	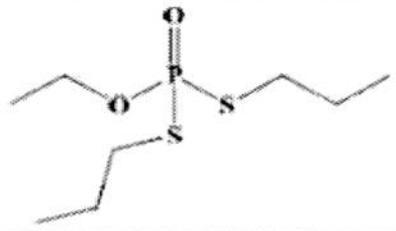	61	26	Ia	N/A	Likely
25	Dichlorvos	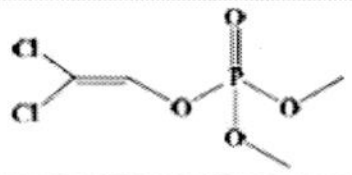	56–80	75–107	Ib	Possible	Suggestive
26	Crotoxyphos	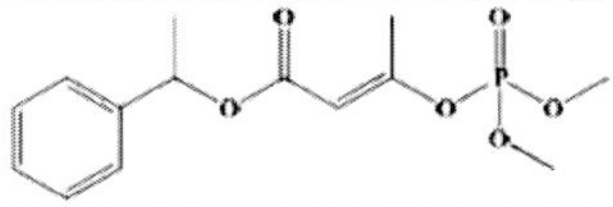	74–110	202–375	N/A	N/A	N/A
27	Naled	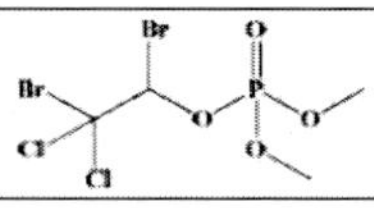	250	800	II	N/A	Unlikely
28	Tribufos	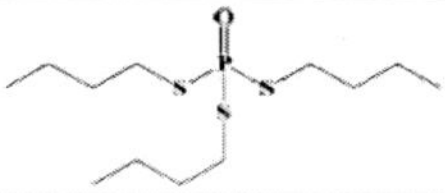	560–630	>2,000	N/A	N/A	Likely (high doses), Not likely (low doses)
29	Trichlorfon	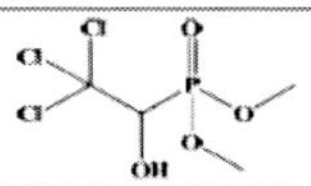	400–800	>2,000	II	Unclassifiable	Likely (high doses), Not likely (low doses)

* Toxic interactions of organophosphorus compounds with any given biological system are dose-related. Their toxicity is expressed in terms of the lethal dose (LD) which will kill 50% of the animal species (LD_{50}). LD_{50} values are generally expressed as an amount per unit weight (e.g., $mg \cdot kg^{-1}$).

§WHO = World Health Organization, acute hazard classify: Ia = extremely hazardous to human health; Ib = highly hazardous; II = moderately hazardous; III = slightly hazardous. ‡ IARC = International Agency for Research on Cancer. † EPA = Environmental Protection Agency.

8.1. Kinetics and Metabolism in Laboratory Animals and Humans [FAO/WHO (1996)]

The methyl parathion is absorbed through the skin and from the respiratory and digestive tracts. Observed differences between its oral and intravenous toxicity are believed to be associated with first-pass effects in the liver. The compound is rapidly excreted; negligible amounts of the labeled dose were present in the blood, tissues, and organs at 48 hours. Conversion of methyl parathion to methyl paraoxon has been shown to occur within minutes after oral administration to rats. Detoxification is achieved by O-demethylation or hydrolysis to p-nitrophenol. In humans, the primary urinary metabolites were p-nitrophenol and dimethyl phosphate.

8.2. The Effects on Experimental Animals and *In Vitro* Test Systems [FAO/WHO (1996)]

Methylparathion is acutely toxic at low doses when administered either orally (LD50 = 4 mg/kg of body weight) or by inhalation (LC50 = 0.13 mg/liter). The compound is slightly irritating to the skin and the eyes but is not a sensitizing agent. WHO has classified methyl parathion as "extremely hazardous." In a 90-day study in mice at dietary levels of 0, 10, 30, or 60 mg/kg, the NOAEL was 10 mg/kg (equivalent to 1.5 mg/kg of body weight per day) based on significant decreases in absolute and relative testicular weights. Cholinesterase activity was not measured in this study.

In a 90-day study in rats, the NOAEL was 2.5 mg/kg (equivalent to 0.12 mg/kg of body weight per day) based on significant decreases in plasma, erythrocyte, and brain cholinesterase activities at 25 mg/kg (equivalent to 1.2 mg/kg of body weight per day).

8.3. The Effects on Humans [FAO/WHO (1996)]

The NOAEL derived from the combined results of several studies conducted in humans, based on the depression of erythrocyte and plasma cholinesterase activities, was 0.3 mg/kg of body weight per day [FAO/WHO (1996)].

Chapter 9

Disposable Methods

The United States, as a signatory to the 1993 Chemical Weapons Convention, has over 30,000 metric tons of chemical warfare agents slated to be destroyed by incineration methods. Currently, incineration is the most efficient method for the disposal of chemical compounds. However, environmental groups have opposed incineration as a disposal method because of the release of byproducts which may pose significant environmental hazards. Transportation of these toxic compounds from storage to disposal sites is both time-consuming and dangerous, considering the threat of potential accidents. Hence, there is a need to monitor insecticide residues in water, soil, plants, and foods. Ozonation, Photochemical Oxidation, Advanced Oxidation Processes (AOPs), and electrochemical methods look to be the most promising methods. Remediation in any of the processes such as microbial, chemical, or photo-degradation or a combination is either introduced or enhanced to reduce the concentration of pesticides in soil or water in a contaminated site. The effective methyl parathion degradation rate by these methods was analyzed and compared to find an eco-friendly, feasible method for the treatment of toxic wastewater with a high concentration of methylparathion. Other disposal methods involve removal, alteration, or contaminant isolation. Usually, for treating contaminated soil, these techniques consist of excavation followed by incineration or containment. Moreover, these remediation technologies are expensive, and most of the time they do not destroy the contaminating compound but rather transfer it from one environment to another. Incineration and chemical treatment are the currently approved methods for the disposal of these materials. However, the destruction of chemical warfare agents by incineration has significant political barriers, and incineration facilities currently used have had discharges of chlorinated organics, heavy metals, and unreacted chemical warfare agents in smoke-stack emissions (Walker and Keasling, 2002).

Given the potential environmental disadvantages of physical and chemical detoxification methods, biodegradation would appear to be a more acceptable alternative. This issue is of such social importance that predictions by the Organization for Economic Cooperation and Development indicated that the worldwide market demand for bioremediation technologies would be very high (Doung et al. 1997).

An ideal treatment method for such pesticide wastes should be non-selective, achieve rapid and complete degradation to inorganic products, and be suitable for small-scale wastes. Various innovative technologies have been proposed for methylparathion treatment. These include the use of UV radiation and hydrogen peroxide (Pignatello and Sun, 1995; Chen et al. 1998). The major disadvantage of these technologies is that they are designed for the decontamination of aqueous solutions with very low active ingredient contents and are not suitable for degrading higher concentrations of unwanted pesticides. Possible methods include biodegradation, chemical reaction, physical entrainment, and incineration. Entrainment is losing acceptance as a final solution for waste disposal. Incineration is prohibitively expensive for diluting aqueous waste because of high-energy requirements and transport costs.

Chapter 10

The Biodegradation/Biotreatment Process and Its Significance

The use of microorganisms (fungi or bacteria), either naturally occurring or introduced, to degrade pollutants is called bioremediation (Pointing, 2001). Microbial metabolism is probably the most important pesticide degradative process in soils (Kearney, 1998) and is the basis for bioremediation, as the degrading microorganisms obtain C, N, or energy from the pesticide molecules (Gan and Koskinen, 1998). The goal of bioremediation is to at least reduce pollutant levels to undetectable, non-toxic, or acceptable levels, i.e. within limits set by regulatory agencies (Pointing, 2001) or ideally completely mineralize organophosphate pollutants to carbon dioxide. From an environmental point of view, this total mineralization is desirable as it represents complete detoxification (Gan and Koskinen, 1998).

As the demand for agricultural products increases, so inevitably does the need for pesticides. Currently, OP compounds are one of the most widely used classes of pesticide in industrialized countries (Shimazu et al. 2001). Despite the rapid degradation of these insecticides in the soil, they can be potentially hazardous as a result of accidental spills, runoff from areas of application, and discharge from pesticide containers and waste (Rani and Lalithakumari, 1994).

Kim et al. (1999), Hertel (1993), and Spain and Gibson (1991) suggested that biodegradation to benign products is an attractive option for the destruction of OP since it utilizes natural processes and offers the potential for less costly treatment. However, biodegradation rates usually are low because the compound being destroyed is toxic or recalcitrant, which causes growth to be slow as well. Numerous decomposition products may be produced so that complete biodegradation often requires a consortium of organisms to metabolize the resulting products. Considerable research has been directed toward the development of alternative processes for the biodegradation of OPs. Among these, Kim et al. (2002) investigated the degradation of coumaphos using a recombinant strain of *Escherichia coli* containing the *opd* gene for OPH. Significantly higher degradation rates were obtained compared to those obtained with the microbial consortium naturally present in coumaphos diphosphate waste.

Bioremediation provides a suitable way to remove contaminants from the environment as, in most cases, OP compounds are mineralized by the microorganisms. Most OP compounds are degraded by microorganisms in the environment as a source of phosphorus carbon or both. Several soil bacteria have been isolated and characterized, which can degrade OP compounds in laboratory cultures and the field. The biochemical and genetic basis of microbial degradation has received considerable attention. Several genes/enzymes, that provide microorganisms with the ability to degrade OP compounds, have been identified and characterized. Some of these genes and enzymes have been engineered for better efficacy. Bacteria capable of complete mineralization are constructed by transferring the complete degradation pathway for specific compounds to one bacterium. Bioremediation has distinct advantages over physicochemical remediation methods as it can be cost-effective and achieve the complete degradation of organic pollutants without collateral destruction of the site material or its indigenous flora and fauna (Timmis and Pieper, 1999). However, the acquisition of biodegradative capabilities by native microorganisms at contaminated sites through natural evolutionary processes such as random mutation often occurs at an unacceptably slow rate, particularly when multiple biodegradation traits are required as is the case with sites contaminated with more than one organic compound (Ang et al. 2005).

Bioremediation overcomes the limitations of traditional methods of hazardous chemical disposal by bringing about the actual destruction of many organic contaminants at a reduced cost. In consequence, over the last 20 years, bioremediation has grown from a virtually unknown technology to a technology that is considered for the remediation of a wide range of contaminating compounds (Hurst et al. 1997). Characteristics of microorganisms such as their small size, ubiquitous distribution, high specificity, surface area, potentially rapid growth rate, and unrivaled enzymatic and nutritional versatility cast them as recycling agents. Moreover, the diversity of inorganic and organic materials present on earth matches the diversity of habitats whose physical and chemical characteristics span wide ranges of pH, temperature, salinity, oxygen tension, redox potential, water potential, etc. This distribution of resources between environments gave origin to a selective evolutionary diversification of microorganisms, resulting in an evolved microbial world capable of exploiting all the naturally occurring metabolic resources on earth (Hurst et al. 1997).

10.1. Microbial Biodegradation of Organophosphate Compounds

The use of microorganisms (fungi or bacteria), either naturally occurring or introduced, to degrade pollutants is called bioremediation (Pointing, 2001). Microbial metabolism is probably the most important pesticide degradative process in soils (Kearney, 1998) and is the basis for bioremediation, as the degrading microorganisms obtain C, N, or energy from the pesticide molecules (Gan and Koskinen, 1998). The goal of bioremediation is to at least reduce pollutant levels to undetectable, non-toxic, or acceptable levels, i.e. within limits set by regulatory agencies (Pointing, 2001) or ideally completely mineralize organophosphate pollutants to carbon dioxide. From an environmental point of view, this total mineralization is desirable as it represents complete detoxification (Gan and Koskinen, 1998).

Figure 7 shows the general approach to the application of bioremediation. This approach can be followed for both in-situ and ex-situ conditions. In-situ bioremediation allows treatment of a large volume of soil at once and it is mostly effective at sites with sandy soils. In-situ bioremediation techniques can vary depending on the method of supplying oxygen or electron donors to the organisms that degrade the contaminants. Three commonly used in-situ methods include bioventing and injection of hydrogen peroxide or oxygen-releasing compound (ORC) for aerobic treatment and injection of HRC for anaerobic treatment. Ex-situ bioremediation involves excavation of the contaminated soil and treatment in a treatment plant located on the site or away from the site. This approach can be faster, easier to control, and used to treat a wider range of contaminants and soil types than the in-situ approach. Ex-situ bioremediation can be implemented as slurry-phase bioremediation, or solid-phase bioremediation (USEPA 1988a; USEPA 1994b).

As the demand for agricultural products increases, so inevitably does the need for pesticides. Currently, OP compounds are one of the most widely used classes of pesticide in industrialized countries (Shimazu et al. 2001). Despite the rapid degradation of these insecticides in the soil, they can be potentially hazardous as a result of accidental spills, runoff from areas of application, and discharge from pesticide containers and waste (Rani and Lalithakumari, 1994). MP is the most hazardous OP insecticide allowed in the United States food supply (Shimazu et al. 2001) and is chemically similar to nerve agents used in chemical warfare (Cho, 2001). MP toxicity is primarily associated with the inhibition of cholinesterase activity and its effects on the nervous system

(Cho, 2001). Although the toxicity of OPs is associated with anticholinesterase activity; some compounds also produce a delayed toxicity termed OP-induced delayed neuropathy (OPIDN). MP can also induce a genotoxic effect in bacterial and mammalian cells. OPIDN is characterized by a delayed onset of pathological symptoms, such as ataxia and muscle paralysis, 7 to 14 days following the exposure (Cho, 2001). Cho (2001) evaluated the neurotoxic effects of MP and its degradation products using an OPH enzyme from genetically engineered *E. coli* to degrade MP and eliminate its neurotoxicity. He found that the metabolites produced were not mutagenic, confirming the genotoxicity of MP and supporting the potential utility of OPH in the bioremediation of MP. Because of these health concerns.

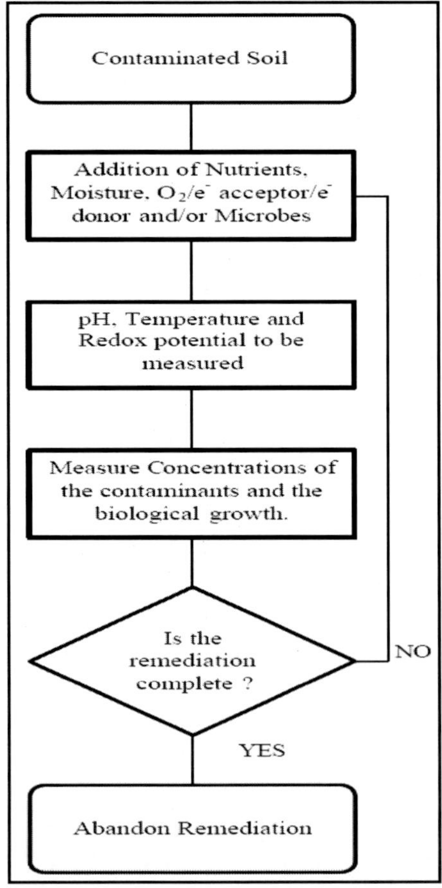

Figure 7. General design approach.

Methylparathion has been classified as a restricted pesticide and its use in most food crops in the United States has been canceled (Garcia et al. 2003). MP is a restricted OP pesticide. The World Health Organization has classified technical MP as a class Ia ‘extremely hazardous’ pesticide in normal use, based on an oral 50% lethal dose in rats of 14 mg/kg. It is highly toxic by inhalation and ingestion and moderately toxic by dermal adsorption as it is readily adsorbed through the skin (Hertel, 1993). MP has adverse effects on many different beneficial insects (Hertel, 1993). It is highly toxic for aquatic invertebrates with most 50 % lethal concentrations ranging from less than 1 μg/L to 40μg/L. Most fish species in both fresh and seawater have 50% lethal concentrations between 6 and 25 mg/L, with a few species substantially more sensitive to MP. MP has been detected at a concentration of 59 μg/kg in the ovaries of spotted sea trout *(Cynoscion nebulosus)* collected in Texas. MP was detected in 34 out of 55 suspected poisoned apiaries examined in Connecticut. Concentrations of MP found in dead bees and brood comb ranged from 0.04 to 5.8 mg/kg (Hertel, 1993). MP is also highly to very highly toxic to birds (Extoxnet, 1996). In 1992, a massive bird kill occurred in Costa Rica after MP was applied by plane in a cotton field (Pesticide News, 1995).

In the United States recently, there have been several important prosecutions involving MP. Over 1,500 homes and businesses in Mississippi and Ohio were sprayed with MP by unlicensed operators. The authorities relocated over 1,100 people to temporary accommodations and clean-up costs could reach US $50 million. On April 30, 1997, the United States EPA canceled the registrations of emulsifiable concentrate formulations (Pesticide News, 1995). Currently, the primary methods approved for the disposal of hazardous wastes are incineration and chemical treatment, but these methods are costly and often create new environmental problems. Kim et al. (1999), Hertel (1993), and Spain and Gibson (1991) suggested that biodegradation to benign products is an attractive option for the destruction of OP since it utilizes natural processes and offers the potential for less costly treatment. However, biodegradation rates usually are low because the compound being destroyed is toxic or recalcitrant, which causes growth to be slow as well. Numerous decomposition products may be produced so that complete biodegradation often requires a consortium of organisms to metabolize the resulting products. Considerable research has been directed toward the development of alternative processes for the biodegradation of OPs. Among these, Kim et al. (2002) investigated the degradation of coumaphos using a recombinant strain of *Escherichia coli* containing the *opd* gene for OPH. Significantly higher degradation rates were obtained compared to those obtained with the microbial consortium naturally present in coumaphos dip waste.

Bioremediation provides a suitable way to remove contaminants from the environment as, in most cases, OP compounds are totally mineralized by the microorganisms. Most OP compounds are degraded by microorganisms in the environment as a source of phosphorus carbon or both. Several soil bacteria have been isolated and characterized, which can degrade OP compounds in laboratory cultures and the field. The biochemical and genetic basis of microbial degradation has received considerable attention. Several genes/enzymes, that provide microorganisms with the ability to degrade OP compounds, have been identified and characterized. Some of these genes and enzymes have been engineered for better efficacy. Bacteria capable of complete mineralization are constructed by transferring the complete degradation pathway for specific compounds to one bacterium.

Bioremediation has distinct advantages over physico-chemical remediation methods as it can be cost-effective and achieve the complete degradation of organic pollutants without collateral destruction of the site material or its indigenous flora and fauna (Timmis et al. 1999). However, the acquisition of biodegradative capabilities by native microorganisms at contaminated sites through natural evolutionary processes such as random mutation often occurs at an unacceptably slow rate, particularly when multiple biodegradation traits are required as is the case with sites contaminated with more than one organic compound (Ang et al. 2005). Bioremediation overcomes the limitations of traditional methods of hazardous chemical disposal by bringing about the actual destruction of many organic contaminants at a reduced cost. In consequence, over the last 20 years, bioremediation has grown from a virtually unknown technology to a technology that is considered for the remediation of a wide range of contaminating compounds (Hurst et al. 1997). Characteristics of microorganisms such as their small size, ubiquitous distribution, high specificity, surface area, potentially rapid growth rate, and unrivaled enzymatic and nutritional versatility cast them as recycling agents. Moreover, the diversity of inorganic and organic materials present on Earth matches the diversity of habitats whose physical and chemical characteristics span wide ranges of pH, temperature, salinity, oxygen tension, redox potential, water potential, etc. This distribution of resources between environments gave origin to a selective evolutionary diversification of microorganisms, resulting in an evolved microbial world capable of exploiting all the naturally occurring metabolic resources on Earth (Hurst et al. 1997).

Biodegradation is a metabolic process that involves the complete breakdown of an organic compound. When this compound is broken down into its inorganic components, the process is referred to as mineralization. Xenobiotic compounds are those having a molecular structure to which microorganisms have not been exposed; therefore, they may be recalcitrant resistant to bioremediation, or not completely degraded. Biodegradability represents the susceptibility of substances to be altered by microbial processes. The alteration may occur by intra-or extracellular enzymatic attack that is essential for the growth of the microorganisms. The attacked substances are used as a source of carbon, energy, nitrogen, or other nutrients or as the final electron acceptor. Cometabolic reactions occur when an enzymatic attack is not necessarily beneficial to the microorganism, i.e., a physiologically useful primary substrate induces the production of enzymes that fortuitously alter the molecular structure of another compound (Hurst et al. 1997). Biodegradation may occur under either aerobic or anaerobic conditions. Under aerobic conditions, oxygen is used as a final electron acceptor; and under anaerobic conditions, microorganisms use compounds such as nitrate (NO_3), sulfate (SO_4^{2-}), or iron (Fe^{3+}) as the final electron acceptor instead of oxygen (Hurst et al. 1997). The rate of biodegradation depends on environmental factors, the number and types of microorganisms present, and the chemical structure of the target compound (Hurst et al. 1997). The process of biodegradation generally involves the breakdown of organic compounds to produce more cell biomass and less complex compounds which are further converted to water and either carbon dioxide or methane. Biodegradation can be intrinsic if the appropriate environmental conditions, nutrients, and microorganisms are present. This is the preferred method because of its lower cost, but enhanced (engineered) bioremediation may be required if the process is not naturally sustained and the rates of degradation are low (Hurst et al. 1997).

10.2. Microbial Metabolism of Organophosphorus Pesticides

Microbial degradation of OP insecticides has been recognized as the most important process controlling their environmental fate (Felsot, 1989). However, the extensive and repeated use of soil-applied OP compounds on certain occasions has led to reduced biological efficacy due to microbial adaptation. This phenomenon was named enhanced or accelerated biodegradation and was attributed to the development of a soil microbial population that was able to rapidly mineralize the OP pesticides. The vulnerability of OP compounds to microbial adaptation has been reported for several compounds including the insecticides parathion (Sethunathan and Yoshida, 1973), diazinon (Sethunathan, 1971; Sethunathan and Pathak 1972), isofenphos (Racke and Coats, 1987), chlorfevinphos (Suett et al. 1996a), phorate (Suett and Jukes, 1997) and the nematicides cadusafos (Karpouzas et al. 2004b), ethoprophos (Smelt et al. 1987; Karpouzas et al. 1999) and fenamiphos (Anderson and Lafuenza, 1992; Stirling et al. 1992; Davis et al. 1993; Singh et al. 2003a; Karpouzas et al. 2004a). Soils exhibiting enhanced biodegradation of organophosphorus compounds, or soils that were heavily contaminated with high concentrations of such compounds, have commonly been used as sources for the isolation of microorganisms with increased capability to rapidly degrade these compounds. The enrichment culture technique in selective mineral salt media, where the OP pesticides act as the sole carbon, nitrogen, or phosphorus source, has been used in the vast majority of cases to isolate such pesticide-degrading microorganisms. In most cases, the isolated microorganisms can utilize the pesticide as a source of a single element (C, N, P, or S). For example, a *Pseudomonas putida* strain was able to use ethoprophos as a carbon source but not as a phosphorus source (Karpouzas et al. 2000b). However, other studies have led to the isolation of microorganisms, which could only co-metabolize and were not able to utilize OP compounds as a source of energy. A list of some OP-degrading microorganisms is presented in Table 3.

Table 3. List of microorganisms capable of degrading organophosphate pesticides

Compound	Microorganisms	References
Parathion	*Bacillus subtilis*	Yasuno et al. (1965)
	Rhizobium spp.	Mick and Dahm (1970)
	Chlorella pyrenoidosa	Zuckerman et al. (1970)
	Flavobacterium sp. ATCC 27551	Sethunathan and Yoshid (1973)
	Bacillus sp and Pseudomonas sp	Siddaramapa et al. (1973)
	Mixed bacterial culture	Munnecke and Hsieh (1974)
	Penicillium waksmani	Rao and Sethunathan (1974)
	Pseudomonas stutzeri,	Daughton and Hsieh (1974)
	Unidentified bacteria	Cook et al. (1978a)
	Pseudomonas sp	Rosenberg and Alexando (1979)
	Pseudomonas diminuta	Serdar et al. (1982)
	Arthrobacter sp., Bacillus sp	Nelson (1982)
	Pseudomonas sp., Xanthomonas sp.	Tchelet et al. (1993)

Compound	Microorganisms	References
Methylpa-rathion	*Bacillus subtilis*	Miyamoto et al. (1966)
	Flavobacterium sp. ATCC 27551	Adhya et al. (1981)
	Trichoderma viride	Baarschers and Heitlandl (1986)
	Pseudomonas sp., Flavobacterium sp	Chaudhry et al. (1988)
	Bacillus sp	Sharmila et al. (1989)
	Bacillus sp	Ou and Sharma (1989)
	Unidentified bacteria	Mishra et al. (1992)
	Pseudomonas putida	Rani and Lalithakumarii (1994)
	Burkholderia sp	Hayatsu et al. (2000)
	Burkholderia cepacia, Bacillus sp	Keprasertsup et al. (2000)
	Plesiomonas sp	Zhongli et al. (2001)
	Pseudomonas sp	Zhongli et al. (2002)
	Pseudomonas sp	Yali et al. (2002)
	Pseudomonas aeruginosa, Pseudomonas aeruginosa mpd-5	Usharani and Lakshmanaperumalsamy,(2016) Usharani Rathinam Krishnaswamy, (2023)
Fenitrothion	*Bacillus subtilis*	Miyamoto et al. (1966)
	Flavobacterium sp. ATCC 27551	Adhya et al. (1981)
	Pseudomonas sp	Adhya et al. (1981)
Diazinon	*Bacillus sp*	Sharmila et al. (1989)
	Arthrobacter sp., Streptomyces sp	Gunner and Zuckerman (1968)
	Flavobacterium sp. ATCC 27551	Sethunathan and Yoshid (1973)
	P. putida	Rosenberg and Alexando (1979)
	Pseudomonas sp	Adhya et al. (1981)
	Arthrobacter sp	Ohshiro et al. (1996)
Chlorpyrifos	*Arthrobacter sp*	Mishra et al. (1992)
	Fungi	Bumpus et al. (1993)
	Micrococcus sp	Guha et al. (1997)
	Arthrobacter sp	Ohshiro et al. (1996)
	Flavobacterium sp. ATCC 27551	Mallick et al. (1999)
	Hypholoma fasciculare, Coriolus versicolor	Bending et al. (2002)
	Enterobacter sp	Singh et al. (2004)
	Alcaligenes faecalis	Yang et al. (2005)
Malathion	*Trichoderma viride*	Matsumura and Boush (1966)
	Aspergillus niger, Penicillium notatum, Rhizoctonia solani	Matsumura and Boush (1966)
	Rhizobium trifolii, R leguminosarum	Mostafa et al. (1972b)
	Artheobacter sp	Walker and Stojanovic (1974)
	Bacterial strains	Paris et al. (1975)
	Bacterial strains	Bourquin (1977)
	Pseudomonas sp	Rosenberg and Alexando (1979)
	Pseudomonas sp	Singh and Seth (1989)
	Aulosira fertilissima, Nostoc muscorum	Subramanian et al. (1994)

Table 3. (Continued)

Compound	Microorganisms	References
	Micrococcus sp	Guha et al. (1997)
	Aspergillus sydowii, A flavus, Fusarium oxysporum	Hasan (1999)
	Pseudomonas sp	Imran et al. (2004)
	Fusarium oxysporum f.sp.Pisi	Kim et al. (2005)
Monocrotophos	*Bacteria (Acinetobacter sp., Nocardia sp., Arthrobacter sp., Pseudomonas sp.)*	Stackhouse (1980)
	Fungi (Alternaria, Alusidium, Gliocladium, Penicillium, Sepedonium	Stackhouse (1980)
	Chlorella vulgaris	Megharaj et al. (1987)
Monocrotophos	*Scenedescusbijugatus, Synechococcus elongates, Nostoc linkia, Phormidium tenue*	
	Aulosira fertilissima, Nostoc uscorum	Subramanian et al. (1994)
	Pseudomonas aeruginosa	Subhas and Singh (2003)
	Clavibacter michiganense	
	A. sydowii	Hasan (1999)
	Pseudomonas aeruginosa	Deshpande et al. (2001)
	A.niger	Liu et al. (2001)
Phorate	*Pseudomonas fluorescens, Thiobacillus thiooxidans*	Ahmed and Casida (1958)
	Streptomyces sp	Gauger et al. (1986)
	Rhizobium sp., Pseudomonas sp., Proteus sp	Bano and Musarrat (2003)
Ethoprophos	*Streptomyces sp*	Gauger et al. (1986)
	P. putidaepi and epii	Karpouzas et al. (2000)
	Sphingomonas paucimobilis, Flavobacterium sp	Karpouzas et al. (20005b)
Cadusafos	*Sphingomonas paucimobilis, Flavobacterium sp*	Karpouzas et al. (20005b)
Isofenphos	*Streptomyces sp*	Gauger et al. (1986)
	Pseudomnas sp	Racke and Coats (1987)
	Arthrobacter sp	Racke and Coats (1988)
	Arthrobacter sp	Ohshiro et al. (1996)
Phosphinothricin	Rhodococcus sp., P. Paucimobilis	Tebbe and Reber (1988)
	Agrobacterium tumefaciens, Alcaligenes sp., Pseudomonas sp., Serratia sp., Enterbacter spp	
Glyphosphate	*Pseudomonas sp. PG2982*	Moore et al. (1983)
	Flavobacterium sp.	Balthazor and Hallas (1986)
	Arthrobacter sp. GLP-1	Pipke et al. (1987)
	Arthrobacter atrocyaneas	Pipke and Amrhein (1988a)
	Arthrobacter sp. GLP-1/Nit-1	Pipke and Amrhein (1988b)

Compound	Microorganisms	References
	Pseudomonas sp. Lbr	Jacob et al. (1991)
	Agrobacterium radiobacter	Mcauliffe et al. (1990)
	Rhizobium meliloti. R. eguminosarum, R. trifolli, R. galega, Agrobacterium rhizogenes, A. tumefaciens	Liu et al. (1991)
	Penicillium cirinum	Zboinska et al. (1992)
	Pseudomonas pseudomallei	Penaloza-Vazquez et al. (1995)
Glyphosphate	*Bacterial strains*	Dick and Quinn (1995)
	Penicillium notatum	Bujacz et al. (1995)
	Streptomyces sp.	Obojska et al. (2001)
	Geobacillus carboxylosilyticus	Obojska et al. (2002)
	Penicillium chromogenum	Klimek et al. (2001)
	Penicillium janthinellum, P. simplicissimum, Mucor sp., Alternaria alternaria	Lipok et al. (2003)
Edifenphos	*Pyricularia oryzae*	Uesugi and Tomizawa (1971)
Pyrazophos	*P. oryzae*	Dewaard (1974)
	Alternaria sydowii, A. flavus, Fusarium oxysporum	Hasan (1999)
Methylparathion	*Fusarium spp, Pseudomonas aeruginosa* *Fusarium sp mpd-1* *Pseudomonas aeruginosa mpd-5*	Usharani and Muthukumar (2013) Usharani and Lakshmanaperumalsamy, (2016) Usharani Krishnaswamy (2021)

To date, many isolated microorganisms could only degrade a limited number of pesticides (Feng et al. 1997; Qiu et al. 2006; Zhang et al. 2006; Kim and Ahn 2009; Li et al. 2009; Wang et al. 2010; Zhang et al. 2010). However, in practice, a variety of different pesticides are used to control different pests during agricultural production. So, it is a mixed pattern of pesticide contamination. There are two ways to resolve the problem. One way is to combine isolates which could degrade different pesticides to eliminate this pollution (Xu et al. 2007; Krishna and Philip 2008). Another way is to isolate the effective bacteria or construct super-bacteria which could degrade several different pesticides. Recently, a genetically engineered microorganism (GEM) was successfully constructed and it could simultaneously degrade methyl parathion (MP) and carbofuran (Jiang et al. 2007). Studies in pure cultures with isolated microorganisms revealed that there are four major reactions involved in OP metabolism: hydrolysis, oxidation, alkylation, and dealkylation. Hydrolysis of the phosphoesteric P–O–C or phosphothiesteric P–S–C bonds present in the OP molecules is considered the initial step in their metabolism. The reduced mammalian toxicity of the hydrolysis products is the main reason for the lack of detailed studies on the subsequent transformations of the produced metabolites by the isolated microorganisms. The approach was to use a multi-step process in which the selected microorganisms were grown rapidly under favorable conditions to achieve a high cell density then used for biodegradation of MP, a model OP compound. The approach was to use a multi-step process in which the selected microorganisms were grown rapidly under favorable conditions to achieve a high cell density then used for biodegradation of MP, a model OP compound. The strategy is based on a rational combination of organisms to provide different catabolic pathways. Ultimately, complete metabolic routes for xenobiotics may be achieved and the formation of toxic metabolites can be avoided. The study focused on the degradation of MP and one of its degradation products, p-nitrophenol (PNP). MP is an OP insecticide that creates potential human risk. PNP and dimethylthiophosphate (DMTP) are the hydrolysis products of MP using the OPH enzyme (Shimazu et al. 2001; Usharani and Muthukuamr, 2013; Usharani and Lakshmanaperumalsamy, 2016; Usharani Rathinam Krishnaswamy, 2021). PNP, like MP, is toxic to human health (Walker and Keasling, 2002).

10.3. Phenyl-Substituted Organophosphates

Parathion (O, O-diethyl-O-p-nitrophenylphosphorothioate) has been one of the most important OP insecticides worldwide. However, its high mammalian toxicity (LD_{50} ¼ 10 mg kg^{-1} body weight) has resulted in its recent withdrawal from the European market (Annex I, Directive 91/414/ EEC). Several studies have documented the involvement of soil microorganisms in the degradation of parathion in soil under both aerobic (Ferris and Lichtenstein, 1980) and anaerobic conditions (Rebby and Sethunathan, 1983). Several studies with parathion-enrichment cultures led to the isolation and characterization of a great variety of bacterial species that were able to hydrolyze parathion (Sethunathan and Yoshida, 1973; Siddaramappa et al. 1973; Adhya et al. 1981). Sethunathan and Yoshida (1973) were the first to report the isolation of a Flavobacterium strain ATCC 27551, which was able to rapidly hydrolyze parathion leading to the accumulation of p-nitrophenol. In a concurrent study, Siddaramappa et al. (1973) isolated two bacteria, a *Bacillus* sp. and a *Pseudomonas* sp. from flooded soil. *Pseudomonas* sp. hydrolyzed parathion and then released nitrite from the p-nitrophenol. On the contrary, *Bacillus* sp. was unable to hydrolyze parathion but was able to use p-nitrophenol as a sole carbon source as soon as it was formed. Munnecke and Hsieh (1974) reported the isolation of a mixed bacterial culture consisting of four *Pseudomonas* sp., a *Xanthomonas* sp., an *Azotomonas* sp., and a *Brevibacterium* sp. which was able to rapidly hydrolyze parathion.

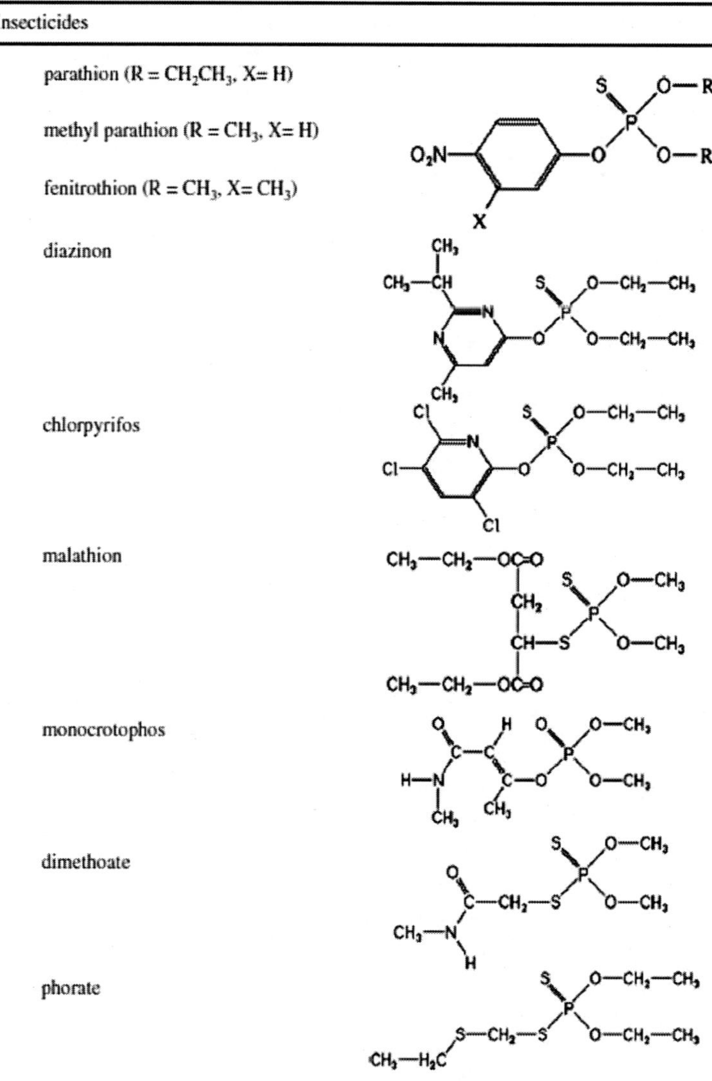

Figure 8a. The molecular structures of organophosphate pesticides include insecticides, nematicides, and fungicides.

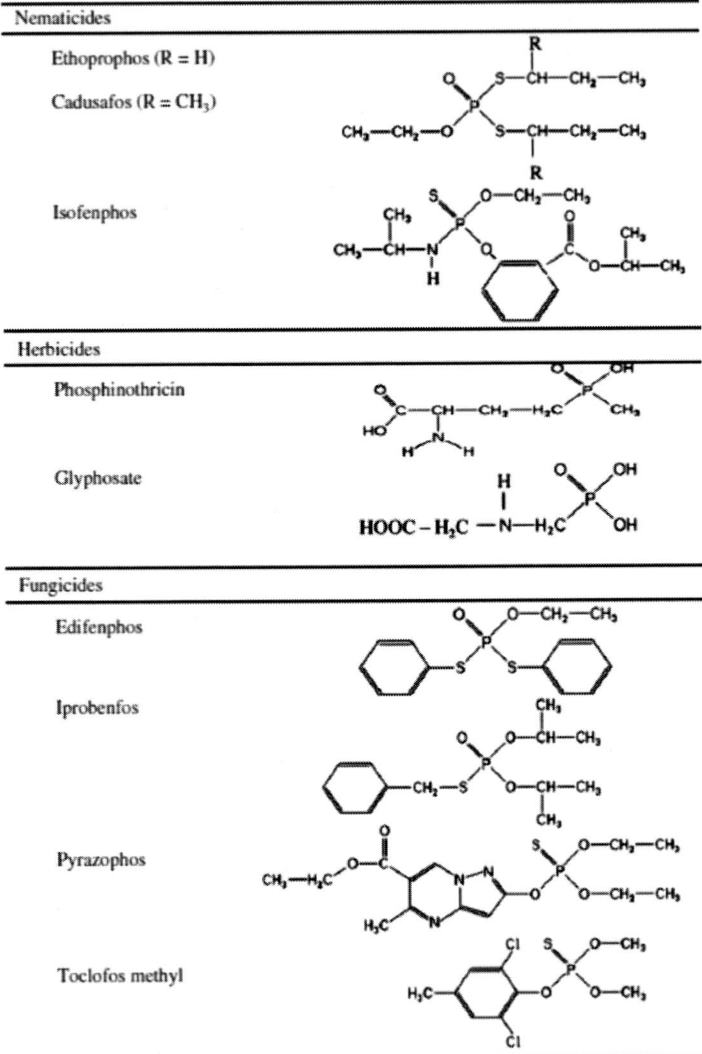

Figure 8b. The molecular structures of organophosphate pesticides include insecticides, nematicides, and fungicides.

Further, metabolic studies revealed that only one of the bacteria was able to metabolize parathion to p-nitrophenol and diethylthiophosphoric acid (DETP) (Usharani and Muthukumar, 2013; Usharani and Lakshmanaperumalsamy, 2016 and Usharani Rathinam Krishnaswamy, 2023). Complementary studies by Munnecke and Hsieh (1976) suggested that parathion degradation by the mixed bacterial culture followed different degradation pathways under aerobic and anaerobic conditions. Under aerobic conditions, the primary pathway involved an initial hydrolysis of parathion-yielding DETP and p-nitrophenol, which was further metabolized to simple non-aromatic products. Under anaerobic conditions, the aromatic nitro group of parathion was reduced to amino parathion, which subsequently undergoes hydrolysis to yield p-aminophenol and DETP. Serdar et al. (1982) isolated from the above-mentioned consortium a *Pseudomonas diminuta* GM strain that possessed a plasmid-mediated hydrolytic mechanism responsible for the hydrolysis of parathion to p-nitrophenol and DETP. Another bacterial consortium isolated from a parathion-adapted soil exhibited a synergistic degrading activity (Daughton and Hsieh, 1977). The bacterial culture was shown to contain a strain of *Pseudomonas stutzeri* capable of rapidly hydrolyzing parathion to DETP and p-nitrophenol; the resultant p-nitrophenol was utilized as a sole carbon and energy source by the other member of the culture, a strain of *Pseudomonas aeruginosa* (Usharani and Lakshmanaperumalsamy, 2016; Usharani Rathinam Krishnaswamy, 2023). The molecular structures of most commonly used organophosphate pesticides including insecticides, nematicides, and fungicides are shown in Figure 8a and 8b.

A study by Nelson (1982) reported the isolation of an *Arthrobacter* and a *Bacillus* strain from a soil sample collected from Israel. Further studies showed that the Arthrobacter strain was able to utilize parathion, but particularly its hydrolytic product, p-nitrophenol, as a sole carbon source, unlike the *Bacillus* strain which was capable of hydrolyzing parathion only in the presence of an extra carbon source. *Pseudomonas* sp and *Xanthomonas* sp were isolated from a pesticide disposal site in northern Israel (Tchelet et al. 1993). The two bacterial strains, although different in the location of their hydrolytic enzymes (intra or extracellular), both degraded parathion in two stages: first p-nitrophenol was released by parathion hydrolysis while in the second stage, p-nitrophenol was degraded. The vast majority of the isolates involved in parathion degradation were able to use parathion or their hydrolysis products as a sole carbon source (Sethunathan and Yoshida, 1973; Siddaramappa et al. 1973; Daughton and Hsieh, 1977). In addition, bacterial isolates were able to utilize the nitrite released from p-nitrophenol hydroxylation as a nitrogen source (Munnecke and Hsieh, 1974). Rosenberg and Alexander (1979) first reported the isolation of two *Pseudomonas* strains that were able to hydrolyze several organophosphates including parathion and to use the ionic cleavage products like DETP as a sole source of phosphorus.

Cook et al. (1978a) isolated an unidentified bacterial strain that utilized dimethyl phosphorothioate, dimethyl phosphorothioate, and their diethyl derivatives as a sole source of phosphorus. These compounds constitute possible primary metabolites produced after hydrolysis of OP compounds. Most of the studies regarding the metabolism of parathion by microorganisms have focused on the primary hydrolytic steps involved in pesticide detoxification, thus overlooking the complete transformation of the resulting metabolites like p-nitrophenol and DETP. Munnecke and Hsieh (1974) first investigated the transformation of p-nitrophenol by a microbial consortium and identified hydroquinone as an early metabolite.

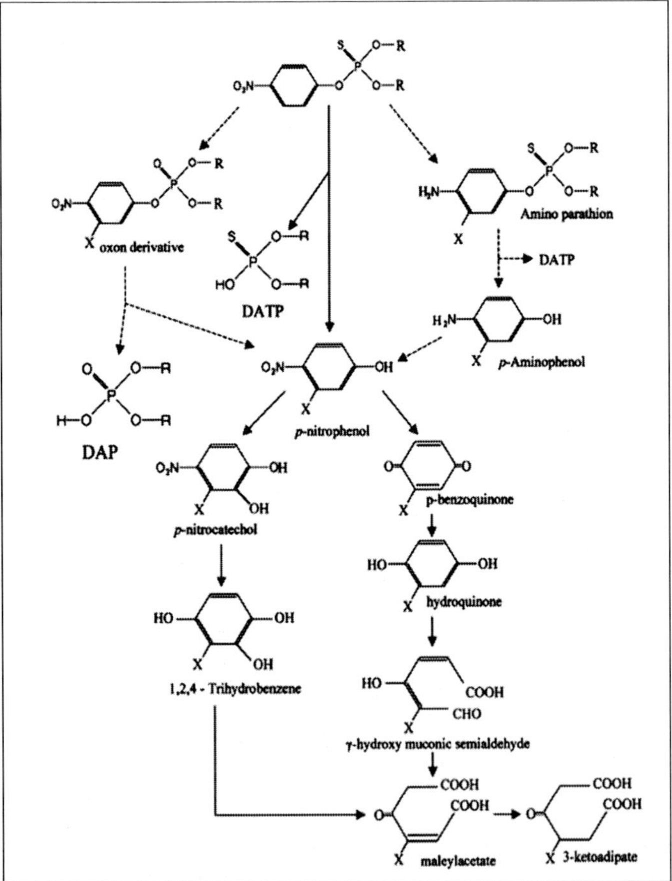

Figure 9. The metabolic pathway of degradation of phenyl-substituted OP insecticides by soil microorganisms.

Where R ¼ CH_3CH_2, X¼ H for parathion, R ¼ CH_3, X ¼ H for methyl parathion, and R ¼ CH_3, X ¼ CH_3 for fenitrothion, DATP (dialkyl thiophosphate). Metabolic steps designated with dashed lines represent minor metabolic pathways.

They then proposed that hydroquinone was hydroxylated to 1, 2, 4-4-benzene-triolrior to ortho ring cleavage. However, Raymond and Alexander (1971) suggested that a *Flavobacterium* sp converted p-nitrophenol into 4-nitrocatechol as the first step before ring fission. Several bacteria belonging to species of *Pseudomonas*, *Moraxella*, *Brevibacterium,* and *Arthrobacter* have been found to metabolize p-nitrophenol with a concurrent release of nitrite (Simpson and Evans, 1953; Spain and Nishino, 1987; Spain and Gibson, 1991; Jain et al. 1994; Ningthoujam, 2005).

Two alternative pathways, for the conversion of p-nitrophenol into a common final product maleylacetate, have been demonstrated. The first pathway was suggested by Spain and Gibson (1991) whereby the formation of p-benzoquinone results in the release of nitrite from p-nitrophenol. Subsequently, hydroquinone is formed and further oxidized by a ring-cleaving dioxygenase to g-hydroxymuconic semi-aldehyde. This is transformed to maleylacetate which is further metabolized to b-ketoadipate.

In the second pathway, an *Arthobacter* sp. and a *Bacillus* sp. hydroxylated p-nitrophenol to produce 4-nitrocatechol which is further oxidized to 1, 2, 4-trihydroxybenzene (THB) with concomitant release of nitrite. THB is further oxidized to maleylacetate which is then converted enzymatically into 3-ketoadipate (Jain et al. 1994; Kadiyala and Spain, 1998). The complete metabolic pathway of parathion degradation by soil microorganisms is presented in Figure 9.

In most studies, hydrolysis of parathion and formation of p-nitrophenol and DETP are the most common initial steps in parathion-microbial metabolism. However, alternative metabolic pathways have also been reported. Munnecke and Hsieh (1976) reported a secondary metabolic pathway of parathion, which involved the oxidation of parathion to paraoxon that was further hydrolyzed to p-nitrophenol and diethylphosphoric acid. Yasuno et al. (1965) found a *Bacillus subtilis* strain that rapidly converted parathion into aminoparathion. Similarly, Mick and Dahm (1970) isolated two *Rhizobium* sp, which rapidly reduced parathion to aminoparathion. Similar degradation pathways have been reported in fungi and algae. A fungus, *Penicillium waksmani*, isolated from an acid sulphate soil, degraded large amounts of parathion to amino parathion and two unidentified polar metabolites (Rao and Sethunathan, 1974). Similarly, Zuckerman et al. (1970)

reported that an alga, *Chlorella pyrenoidosa*, was able to metabolize parathion to aminoparathion. Methyl parathion (O, O-dimethyl O-(p-nitrophenyl) phosphorothioate) and Fenitrothion (O, O-dimethyl O-4-nitro-m-tolyl-phosphorothioate) are still widely used due to their relatively lower mammalian toxicity compared to their analog, parathion. Several microorganisms that were isolated from parathion enrichments have also been tested for their ability to metabolize methyl parathion and fenitrothion (Rosenberg and Alexander, 1979; Adhya et al. 1981). Only a few studies have focused on the isolation of methyl parathion-degrading microorganisms and the investigation of the associated metabolic pathways. Chaudhry et al. (1988) isolated a *Pseudomonas* sp. and a *Flavobacterium* sp. from soil collected from a farmyard previously treated with methyl parathion. *Pseudomonas* sp. hydrolyzed the pesticide to p-nitrophenol but required glucose or another carbon source for growth, unlike *Flavobacterium* sp., which was able to metabolize p-nitrophenol by releasing nitrite which was used by the bacterium as a nitrogen source. *Bacillus* sp. isolated by Sharmila et al. (1989) was able to hydrolyze methyl parathion, parathion, and fenitrothion in the presence of different concentrations of yeast extract. Particularly noteworthy is the finding that the same bacterium affected both, nitro group reduction and hydrolysis of methyl parathion depending on the concentration of yeast in the liquid medium. Degradation of methyl parathion proceeded via hydrolysis to p-nitrophenol and DMTP (dimethyl thiophosphate) in the presence of a concentration (w/v) of yeast extract at 0.5%, by both hydrolysis and nitro group reduction at 0.1 and 0.25%, and exclusively by nitro group reduction at 0.05%. In contrast, degradation of fenitrothion by *Bacillus* sp. proceeded via hydrolysis regardless of the concentration of yeast. Ou and Sharma (1989) isolated another *Bacillus* strain that utilized methyl parathion as a carbon and energy source. However, Rani and Lalithakumari (1994) first reported the isolation of a *Pseudomonas putida* which utilized methyl parathion as the sole carbon and/or phosphorus source. In addition, *P. putida* also utilized the metabolic products derived from the degradation of methyl parathion, such as p-nitrophenol, hydroquinone, and 1,2,4-benzenetriol, as carbon sources. Subsequently, the later metabolite was further transformed by the *P. putida* strain to maleylacetate following the same metabolic pathways as described before for parathion. Keprasertsup et al. (2001) reported *Burkholderia cepacia* was isolated from a methyl parathion-treated site in Thailand. This isolate was able to rapidly degrade methyl parathion and p-nitrophenol and utilize them as sole sources of carbon. A *Bacillus* sp. and two unidentified pure cultures that were isolated in the same study were able to degrade commercial-grade methyl parathion and not analytical-grade methyl parathion. This finding indicates that the pesticide was co-metabolized by these bacteria that were grown on other organic compounds present in the pesticide formulation.

In a concurrent study, Zhongli et al. (2001) isolated a *Plesiomonas* sp., which was able to hydrolyze 200 mg of methyl parathion to p-nitrophenol within 15 min, but it was unable to further transform p-nitrophenol that was accumulated in the medium. More recently, the same group isolated a *Pseudomonas* sp. strain p3 which was able to utilize methyl parathion as a sole carbon and nitrogen source (Zhongli et al. 2002). In a subsequent study, another *Pseudomonas* sp. was isolated from polluted soils around a Chinese pesticide factory, which effected complete degradation of methyl parathion by using the pesticide as a sole source of carbon and nitrogen (Yali et al. 2002). Mishra et al. (1992) reported the isolation of two bacterial isolates that rapidly hydrolyzed methyl parathion, parathion, and fenitrothion to p-nitrophenol and which was further metabolized with concomitant release of nitrite. In contrast, 3-methyl-4-nitrophenol, the hydrolysis product of fenitrothion, was not further metabolized by the isolated bacteria. Similar studies by Adhya et al. (1981) reported that the parathion-degrading *Flavobacterium* strain ATCC 27551 hydrolyzed fenitrothion. However, Hayatsu et al. (2000) isolated *Burkholderia* sp. NF100 which utilized both fenitrothion and methyl parathion as carbon sources. The metabolic pathway of methyl parathion and fenitrothion by *Burkholderia* sp. NF100 involved an initial hydrolysis of p-nitrophenol and 3-methyl-4-nitrophenol, respectively. These products were further oxidized to hydroquinone and methyl hydroquinone, as has been described before. Evidence for the microbial degradation of fenitrothion and its oxon analog, fenitrooxon, by fungal isolates was provided by Baarschers and Heitland (1986), who found that the fungus Trichoderma viride could hydrolyze both compounds to 3-methyl-4-nitrophenol which was then further degraded by co-metabolic reactions. Previous studies by Miyamoto et al. (1966) observed a different degradation pathway for fenitrothion by a *B. subtilis* strain. The major metabolite, accounting for 65% of the added insecticide, was amino fenitrothion; other minor metabolites detected were DMTP and dimethyl fenitrothion. The amino fenitrothion was then slowly transformed into dimethyl amino fenitrothion. Methyl parathion was metabolized by *B. subtilis* in the same way as fenitrothion but twice as fast.

10.4. Parathion Degradation by Microorganisms

Several soil bacteria are capable of degrading parathion. They synthesize parathion hydrolase, which is encoded by the opd (for "organophosphate degradation") gene and which carries out the first step in parathion degradation leading to the production of p-nitrophenol and diethyl phosphoric acid (Munnecke et al. 1976). Microbial enzymes such as parathion hydrolase are thought to play a significant role in the degradation of parathion and related organophosphorus in the soil (Siddaramappa et al. 1973) and this has stimulated studies of the organisms that are capable of synthesizing such enzymes (Karns et al. 1987; Mulbry and Karns, 1989a,b; Mulbry and Kearney, 1991).

Organophosphates such as parathion, it has been relatively easy to isolate degrading bacteria: two different strains, *Flavobacterium* sp. Strain ATCC 27551 and *Pseudomonas diminuta* strain GM, have been isolated from soils in the Philippines and the US, respectively (Sethunathan and Yoshida, 1973 Serdar et al. 1982). In addition, studies by Rani and Lalithakumari (1994) found that a strain of *Pseudomonas putida* could hydrolyze methyl parathion and use p-nitrophenol as a sole source of carbon. Parathion hydrolyze-producing *Flavobacterium* sp was isolated in 1972 from a rice field in the Philippines (Sethunathan and Yoshida, 1973). In 1982 a strain of *Pseudomonas diminuta* that could utilize parathion as a carbon source was isolated in Texas (Serdar et al. 1982), and in 1995 an organism identified as *Flavobacterium bakistinum*, which grew methyl parathion as the sole carbon source, was isolated from agricultural soils close to Anantapur in Andhra Pradesh, India (Somara and Siddavattam, 1995). In each of these organisms the opd gene was found to be located on a large plasmid (Mulbry et al, 1986, Harper et al. 1988 Somara et al. 2002). Microbial enzymes have been shown to hydrolyze parathion under controlled conditions. Munnecke and his co-worker in 1975 first reported the ability of parathion hydrolase, an organophosphorus ester-hydrolyzing enzyme isolated from a mixed microbial culture, to hydrolyze chlorpyrifos. The parathion hydrolase prevents a 66-kb plasmid in *Pseudomonas diminuta* isolated from an enrichment culture (Serder et al. 1982). The gene encoding this hydrolase has been cloned and introduced into various bacterial strains (Steriert et al. 1989), *Escherichia coli* (Serder et al. 1989), and insect cells (Dumas et al. 1989). The gene encoding a parathion hydrolase in Flavobacterium sp was found to be a 43kb-plasmid (Mulbry et al. 1986). Munnecke and Hsieech (1976) have studied the degradation of parathion by mixed microbial cultures and have found the major metabolites to be p-nitrophenol (PNP) and diethyl thiophosphoric acid. PNP can be further utilized as a source of carbon and energy by other microorganisms. Recently, *Pseudomonas putida* was reported to metabolize PNP to hydroquinone and 1,2,4- benzenetriol; even finally the ring compound was cleaved by benzenetriol oxygenase to maleyl acetate (Rani and Lalithakumari, 1994). DNA-DNA hybridization experiments using the parathion hydrolase gene (termed opd for organophosphate degradation) from *Pseudomonas diminuta* and *Flavobacterium* indicated that the genes from these two sources were very similar, if not identical. Restriction mapping of cloned DNA fragments from both organisms also suggests that the DNA encoding the gene itself is very similar in these two organisms. Mulberry et al. (1987) observed that the parathion hydrolase in two temporally, geographically, and biologically different isolates of bacteria was encoded by identical genes carried on non-identical plasmids suggesting that the gene may be part of a mobile genetic element or transposon. The hydrolase-producing *Flavobacterium* strain has been used in a pilot-scale system to detoxify waste containing high concentrations of the organophosphate insecticide coumaphos (Karns et al. 1987).

Recombinant plasmids containing the opd gene were transferred into a gram-negative host *E coli*. However, the opd gene was very poorly expressed under the control of its promoter (Mulbry and Karns, 1989a; Mulbry and Karns, 1989b; Cork and Crueger, 1991). Increased expression of the "opd gene" was achieved when cloned downstream to an *E coli* promoter sequence (such as the lac promoter). Expression in gram-positive bacteria has also been reported (Steiert et al. 1989). In fungi, Supak (1990) reported transfer and low expression of the opd gene in *Gliocladium virens*. This fungus being a biocontrol agent against many soil-borne pathogenic fungi holds great potential for twin functions of biocontrol and bioremedial agents. Dave et al (1994) used a 9.4-kb plasmid, pCLI, to transform *Gliocladium virens*. Plasmid pCLI was derived from pJS294 by placing the fungal promoter (promI) from *Cochliobolus* heterostrophus upstream and the trpC terminator from *Aspergillus nidulans* downstream of the opd gene. Consequently, integration occurred at nonspecifically multiplied sites and produced a processed enzyme. The organophosphate hydrolase (OPH) activity was found to be proportional to biomass production. Anchoring OPH onto the surface of *E coli* using an LPP-OmpA (46-159) fusion system, Richins et al. (1997) developed a unique approach for organophosphorus pesticide detoxification. It was anticipated that immobilization of these live biocatalysts onto solid support could provide an attractive and economical means for pesticide detoxification as compared to immobilized enzymes or whole cells. Although recombinant strains effectively degraded parathion they were relatively unstable. To overcome this problem, a new plasmid was constructed to express OPH onto the cell surface under the control of a tightly regulated lac promoter. Production of active OPH onto the cell surface was highly host-specific; a high rate of parathion degradation was observed from strains JM 105 and XL1-blue (Kaneva et al. 1998). The important advantages of microbial degradation and mineralization of pesticides are the removal of these toxicants from the natural environment, maintenance of natural equilibrium, and decreasing the hazardous effects on the health of animals and mankind.

10.5. Methyl Parathion Degradation by Microorganism

In nature, microorganisms have evolved degradative pathways as a result of continuous or repeated exposure to xenobiotic chemicals such as OPs. However, because of the enhancement of microbial degradation of many chemicals, the efficacy of several pesticides including OPs has been reduced (Rani and Lalithakumari, 1994). In particular, loss of insecticidal activities has been reported in soils that have received continuous applications, resulting in the enhanced degradation of these compounds by soil microorganisms (Chaudhry et al. 1988). Ramanathan and Lalithakumari (1996) showed that some microorganisms can use methylparathion (MP) as a carbon source. Studies on a natural microbial community showed that concentrations of MP up to 5 mg/L increased biomass and reproductive activity, but higher soil concentrations were found to reduce microbial reductive potential. An MP-positive effect was observed in bacteria and actinomycetes, while fungi and yeasts were less able to utilize the compound (Hertel, 1993). Adhya et al. (1981) reported the rapid hydrolysis of MP and fenitrothion by a *Flavobacterium* sp. Sharmila et al. (1989) stated that MP is susceptible to degradation by hydrolysis to PNP and DMTP in soil and water environments, by nitro group reduction to methylamino parathion, or both. Hydrolysis is the leading pathway in non-flooded soil, while MP is degraded essentially by nitro group reduction in predominantly anaerobic ecosystems such as flooded soil. In a few instances, hydrolysis is the major or only pathway of MP degradation in soils even under flooded conditions (Sharmila et al. 1989). Chaudhry et al. (1988) isolated two mixed bacterial cultures by soil enrichment that was capable of utilizing MP as a sole source of carbon. The results from their study indicated that mixed cultures are more stable in retaining their ability to completely degrade MP than isolated bacteria. Mishra et al. (1992) isolated a *Flavobacterium* sp. (ATCC 27551) from diazinon-treated rice fields that could use MP as a sole source of carbon. A *Bacillus* sp. was isolated from MP-treated flooded soil but required yeast extract for the degradation of MP in a mineral salt medium. Rani and Lalithakumari (1994) and Ramanathan and Lalithakumari (1999) investigated the degradation of MP by *Pseudomonas putida* and *Pseudomonas* sp. A3, respectively, finding that MP is utilized as both carbon and phosphorus sources by both bacteria. *Pseudomonas aeruginosa mpd-5* predominantly used methylparathion as a sole carbon and phosphorus source as well as biotransformed into nontoxic end products with intermediates paraoxon and para nitrophenol (Usharani and Muthukumar, 2013; Usharani and Lakshmanaperumalswamy, 2016; Usharani Rathinam Krishnaswamy, 2023) *Pseudomonas putida* also utilized one of the intermediates of MP degradation, PNP, as a sole carbon source. Ramanathan and Lalithakumari (1999) reported that *Pseudomonas* sp. A3 completely removed an initial concentration of 0.25 gL^{-1} MP from the medium in 40 h; however, when the initial concentration was 1 gL^{-1}, only 45% of the MP was degraded in 56 h, which gave a degradation rate of 0.51 mol/L·min.

An alternative approach to the degradation of OPs is to use enzymes produced by these microorganisms. Shimazu et al. (2001) isolated an organophosphate hydrolase from soil microorganisms that have been shown to hydrolyze a wide range of OP pesticides. Enzymatic hydrolysis of MP leads to the formation of DMTP and PNP as byproducts, thus reducing its toxicity by 120-fold (Shimazu et al. 2001). The gene sequence of OP pesticide hydrolase genes from *P. diminuta* GM and *Flavobacterium* sp. Strain ATCC 27551 has been studied. The two genes were found to share 86.3% identity on the nucleic acid level within the coding region (Zhongli et al. 2001). Bujacz et al. (1995) studied the utilization of organophosphates by a wild-type strain of *Penicillium notatum*. Results indicated that the fungus may play an important role in biodegradation of organophosphates. The organophosphorus insecticides including chlorpyrifos, malathion, parathion, etc. are moderately persistent. *Alternaria alternata*, *Cephalosporium* sp., *Cladosporium cladosporioides*, *Cladorrhinum brunnescens*, *Fusarium* sp., *Rhizoctonia solani*, and *Trichoderma viride*, reveal the degradation of chlorpyrifos in liquid culture (Singh, 2006; Usharani and Muthukumar, 2013; Usharani Rathinam Krishnaswamy, 2023). A fungus capable of utilizing carbofuran as a sole carbon and energy source was characterized and identified as being a member of the genus *Gliocladium* (Slaoui et. al., 2007) and *Fusarium* sp mpd-1 (Usharani and Muthukumar, 2013; Usharani Rathinam Krishnaswamy,

2023). *Fusarium* sp mpd-1 predominantly used methylparathion as a sole carbon and phosphorus source as well as biotransformed into nontoxic end products with intermediates paraoxon and para nitrophenol (Usharani and Muthukumar, 2013; Usharani and Lakshmanaperumalswamy, 2016; Usharani Rathinam Krishnaswamy, 2023) *Aspergillus niger* has also shown high degrading potential against several pesticides including organophosphates (Liu et al. 2001), Chlorpyrifos in pure cultures and soil could be degraded by *Aspergillus* sp. Y and *Trichoderma* Pres. ex Fr Y, *Alcaligenes faecalis*DSP3, *Fusarium* LK and *Bacillus latersprorus* DSP (Fang et. al., 2008). Thus, environmental organophosphate contaminants are a big threat to the environment but recent reports of the capability of diversified fungi to utilize contaminants in different biochemical processes of primary and secondary processes provide some hope for the development of in-situ bioremediation process to get successful results.

10.6. p-Nitrophenol Degradation by Microorganism

Biodegradation of nitrophenols has been well documented as well as their respective metabolic pathways; however, little information is available on the kinetics of PNP degradation and the effects of metabolic intermediates on its degradation. The toxicity and the poor biodegradability exhibited by these compounds are the main disadvantages in the application of bioremediation for their treatment (Bhushan et al. 2000). Bhushan et al. (2000) studied the rate of PNP degradation by different PNP-degrading bacteria through the accumulation of metabolic intermediates and their effects on PNP degradation. Bhushan et al. (2000) also, investigated PNP degradation kinetics by *Ralstonia* sp. SJ98, *Arthrobacter protophormiae* RKJ100, and *Burkholderia cepacia* RKJ200. They found that all three bacteria utilized PNP as a sole carbon and nitrogen source and that *Ralstonia* sp. SJ98 showed the highest values for both substrate saturation constant and maximum degradation rate. Chauhan et al. (2000) discovered that *Arthrobacter protophormiae* strain RKJ100 was capable of utilizing PNP as well as 4-nitrocatechol (NC) as carbon and nitrogen sources. Although the nitro group of PNP enhances the resistance of the aromatic ring to biodegradation, bacterial strains capable of degrading PNP have been isolated (Leung et al. 1997 Chauhan et al. 2000). Usually, aromatic compounds are activated for further reactions by the introduction of two hydroxyl groups, either in *ortho-* or *para*-position to one another, which in the case of hydrophobic aromatics is usually achieved by multi-component dioxygenases (Pieper and Reineke, 2000). Nitroaromatic compounds can be degraded by aerobic bacteria through a variety of oxidative or reductive pathways. Nitroaromatic compound degradation pathways involve either the removal or the reduction of the nitro group and the conversion of the resulting molecule into a substrate for oxidative ring fission (Hurst et al. 1997). Thus, two major pathways have been proposed for the aerobic bacterial degradation of PNP.

Aerobic degradation of PNP can be initiated by the formation of either hydroquinone (HQ) or NC (Chauhan et al. 2000). The HQ degradation pathway, found in *Moraxella* sp. and *Pseudomonas putida* involves the formation of HQ by an initial monooxygenase-catalyzed reaction where the nitro group of PNP is replaced by an -OH group with the release of nitrite. HQ is then converted to beta-ketoadipic acid via gamma-hydroxymuconic semialdehyde and maleylacetic acid (Spain and Gibson, 1991). The NC degradation pathway, found in *Arthrobacter* sp., *Flavobacterium* sp., and *Rhodococcus* sp. involves the formation of NC by a 2- monooxygenase followed by removal of the nitro group as nitrite via a NC-4- monooxygenase (Leung et al. 1997). However, Chauhan et al. (2000) who studied the degradation of PNP and NC by *Arthrobacter protophormiae* RKJ100 found that PNP degradation proceeded with the formation of p-benzoquinone (BQ) and HQ and was further degraded via the beta-ketoadipate pathway. Among the bacterial strains able to degrade PNP, Barik et al. (1978) isolated a particular strain of *Pseudomonas* sp., *Burkholderia cepacia* ATCC 29354, from parathion-amended flooded soil. The organism degraded PNP via a pathway involving ring hydroxylation with NC as the stable intermediate. No growth of the bacterium was observed despite the degradation of PNP. Doung et al. (1997) investigated the stoichiometry and kinetics of the hydroxylation of PNP to NC by the same bacterial strain in a batch system. It was noted that PNP removal did not correspond to an accumulation of NC, implicating another relatively short-lived biochemical pathway for the breakdown of PNP. Spain et al. (1984) studied the acclimation of microbial communities exposed to PNP in laboratory test systems and in a freshwater pond finding that degradation of PNP required acclimation in both cases. Zaidi and Mehta (1995) studied the degradation of PNP and the effect of a supplementary carbon source on PNP degradation by *Corynebacterium* Z4, *Pseudomonas* MS, and *Pseudomonas* GR in lake and industrial wastewater. Their results suggested that a supplementary carbon source increases or decreases the rate of degradation depending on the bacteria. Several studies of immobilized cells for PNP degradation have been carried out. Errampalli et al. (1999) observed the survival of free and immobilized cells of *Moraxella* sp. G21 in sterile and non-sterile soil contaminated with PNP. Cells survived for a month and immobilized *Moraxella* sp. G21 cells and free cells produced similar PNP mineralization values. Bhatti et al. (2002) studied the use of nonwovens as a cell retainer for PNP degradation in a continuous flow system. It was found that cultivation of the PNP-degrading cells before attachment was necessary to achieve degradation and that the use of nonwoven supports allowed consistent high-rate PNP degradation.

In another study, plasmid-harboring operons encoding enzymes for PNP transformation to p-ketoadipate were transformed into *P. putida* allowing the organism to use 0.5 mM PNP as a carbon and energy source (Walker and Keasling, 2002). Leung et al. (1997) investigated the transformation of PNP by *S. chlorophenolicum* UG30 finding that NC is an intermediate product. This suggested that PCP-4-monooxygenase does not catalyze the first step of PNP transformation by this strain (Leung et al. 1997). To verify this, Leung et al. (1999) cloned the gene encoding PCP-4-monooxygenase from *S. chlorophenolicum* UG30 to study its potential role in PNP degradation, confirming that PCP-4-monooxygenase is not the primary enzyme in the initial step of PNP metabolism by *S. chlorophenolicum* UG30 but indicating that the enzyme may be involved in the second step of PNP degradation. Cassidy et al. (1999) examined the ability of several PCP-mineralizing strains (*S. chlorophenolicum* UG30, and *S. Chlorophenolicum* ATCC 39723 and RA2) to metabolize PNP. All of the strains degraded PNP with nitrite release. This suggests that UG30 degrades PNP by the NC pathway. Dechlorination of PCP by bacteria has been extensively studied (Leung et al. 1999 Wang et al. 2002). Xu et al. (1999) studied the effect of metals and found that only Fe^{2+} supported enzyme activity. Topp et al. (1988) and Chang and Su (2003) studied the effect of carbon source on PCP degradation indicating that supplementary carbon decreased acclimation time and facilitated PCP metabolism by glucose or glutamate-grown *Flavobacterium* cells and that, as expected, yields were higher when cells received more carbon source. Leung et al. (1997) studied the transformation of PNP by PCP-degrading *S. chlorophenolicum* UG30 and *S. chlorophenolicum* strains RA2 and ATCC 39723 in either mineral salts-glutamate or mineral salts-glucose medium. It was observed that mineralization of PNP only occurred in the mineral salts-glucose medium. Particular studies with UG30 showed that this strain did not transform or mineralize PNP in a growth medium lacking glucose or glutamate and that pre-exposure of UG30 cells to PNP eliminated an initial lag phase in PNP transformation (Leung et al. 1997). Crawford and Ederer (1999) described the genus *Sphingomonas* as Gram-negative, non-spore-forming, aerobic, yellow-pigmented, straight rods. It was also mentioned that the yellow pigmentation in *Sphingomonas* is due to the carotenoid nostoxanthin. Also, members of this genus are characterized by the presence of sphingoglycolipids. *Sphingomonas* species also contain octaecanoic acid, 2-hydroxymyristic acid, cis-9-hexadecenoic acid, and hexadecanoic acid as major fatty acids, the ubiquinone Q10 as the major respiratory quinone, and DNA containing 62-67% GC. Organisms in the genus *Sphingomonas* secrete gellan-related polysaccharides. Finally, Crawford and Ederer (1999) mentioned that this genus is best grown on dilute media. The presence of sphingolipids and octadecenoate is indicative of the genus *Sphingomonas* (Crawford and Ederer, 1999). Although many of the previous studies on the degradation of MP and PNP provide excellent information about the bacteria involved and degradation pathways, none have directly studied the hydrolysis kinetics of MP by recombinant *Escherichia coli* or have fully assessed the degradation of PNP by *Burkholderia cepacia* and *Sphingobium chlorophenolicum.* They studied the effect of biomass and substrate concentrations on the degradation of MP by recombinant *Escherichia coli* and have established conditions for the degradation of PNP by *Burkholderia cepacia* and *Sphingobium chlorophenolicum.* Hydrolysis of MP can be accomplished using a dense, non-growing population of a recombinant OPH+ *E. coli* strain that functions without the addition of nutrients required for growth. Degradation of PNP can be accomplished using *Burkholderia cepacia* deposited as *Pseudomonas* sp. (ATCC 29354) and *Sphingobium chlorophenolicum* deposited as *Flavobacterium* sp. (ATCC 53874).

10.7. Microbiological Transformation of Methylparathion and Its Metabolites

Microorganisms degrading xenobiotic chemicals have elaborate enzyme systems. The biodegradation of organophosphates involves the activities of phosphatases, esterases, hydrolases, and oxygenases. Munnecke (1976), working with parathion, diazinon, chlorpyriphos, and methyl parathion showed enzymatic hydrolysis occurring at the aryl P-O bond. Microbial degradation through hydrolysis of P–O aryl and P–O aryl bonds is considered the most significant step in the detoxification of organophosphorus compounds. The hydrolase responsible for catalyzing this reaction, organophosphate hydrolase has been isolated from soil microorganisms. The bacterial phosphotriesterases were reported to be the most promiscuous of all enzymes (Scott et al. 2008). Generally, they have a broad substrate range, being able to hydrolyze several related compounds. In addition to the hydrolysis of P–O bonds in phosphotriesters, they also could catalyze the hydrolysis of P–S bonds (Lai et al. 1995), P–F bonds (Dumas et al. 1990; Watkins et al. 1997), P–CN bonds (Raveh et al. 1992), and C–O bonds in esters and lactones (Roodveldt and Tawfik 2005b; Roodveldt and Tawfik 2005a). Biodegradation of organophosphorus pesticides by organophosphorus hydrolase has been reported by Richins et al. (1997). Dumas et al. (1989) reported the hydrolysis of parathion, diazinon, fensulfothion, and coumaphos by the enzyme phosphotriesterase. Singh et al. (2004) also reported higher phosphotriesterase activity in *Enterobacter* strain B-14, which hydrolyzed 35 mgL^{-1} of chlorpyriphos within 24 hr when inoculated with 10^6 cells/mL to the medium. Singh and Subhas (2002) reported the enzyme phosphotriesterase to be responsible for monocrotophos degradation by the two bacterial isolates *Pseudomonas aeruginosa* F10B and *Clostridium michiganense* subsp *insidiosum* SBL11. They reported that the gene responsible for the production of phosphotriesterase is plasmid-borne. The cell-free extract of *Flavobacterium balustinum* was found to hydrolyze a variety of organophosphorus pesticides namely fenitrothion, quinalphos, and monocrotophos (Sureshkumar et al. 1998) (Table 4 and Table 5). Degradation strategies exhibited by microorganisms include cometabolism - the biotransformation of a molecule coincidental to the normal metabolic functions of the microbe; catabolism- the utilization of the molecule as a nutritive or energy source; and extracellular enzymes (phosphatases, amidases, and laccases), secreted into the soil, which can act on the molecule as a substrate. Three basic types of reactions can occur: degradation, conjugation, and rearrangements, all of which can be microbially mediated. The complete degradation of a chemical in the soil to carbon dioxide and water involves many different types of reactions. Microorganisms are key players in determining the environmental fate of novel compounds because they can be used as carbon and energy sources by microorganisms. The microbial hydrolysis of organophosphorus pesticides by *Pseudomonas putida* and *Flavobacterium* sp. is carried by membrane-bound enzymes. Genes of these enzymes are also found on plasmids (Allard et. al., 1997).

The organophosphates are mainly detoxified through oxidation and hydrolysis. The two main enzyme groups involved in the hydrolysis of these compounds are phosphotriesterases (PTEs) and carboxylesterases (CbEs).

Phosphotriesterases, being of unknown physiological role, hydrolyze (in some cases stereo-specifically) organophosphorus insecticides (OP). PTEs have been found in a multitude of species, from mammals to bacteria. It offers application in the therapy of organophosphorus poisonings, biodegradation, and bioremediation of organophosphates.

Table 4. Enzymes responsible for microbial degradation of organophosphates

S. No	Enzymes	References
1	Esterases	Dumas *et al* (1989)
2	Lyases	Munneke (1977)
3	Phosphatases	Bourquin (1977)

The organophosphate hydrolases (OPH) enzyme is effective in degrading a range of organophosphate esters (Mulbry et al. 1986 Serder et al. 1989). Recombinants OPH have been demonstrated to be equally effective in hydrolyzing various organophosphorus pesticides (Phillips et al. 1990; Xu et al. 1996). Bacterial phosphotriesterase (PTE), also known as OPH is a highly efficient hydrolytic enzyme that can hydrolyze a broad range of organophosphates (Hong et al. 1996). PTE catalyzes the cleavage of P - O, P - F (or) P – S bonds in these organophosphates. Although PTE is thought to have only evolved within the last 50 years its hydrolytic ability is truly remarkable (Ang et al. 2005). Hydrolysis of phosphate esters, catalyzed by esterase, is an important mechanism for organophosphate pesticides. For example; an esterase hydrolyzed the P- O- C linkage in parathion after a nitro-reduction, which leads to the formation of p-aminophenol (Zhang and Bennett, 2005).

Microbial degradation of organophosphorus insecticides has been described previously, eg: the degradation of parathion and diazinon by *Flavobacterium* sp (Sethunathan and Yoshida, 1973) and by *Pseudomonas* sp (Siddaramappa et al. 1973). Recently, the biodegradation of chlorpyrifos in pure cultures and soil by *Flavobacterium* sp and *Arthrobacter* sp was investigated by Mallick et al. (1999) (Table 5). Microorganisms play a key role in the degradation of organophosphate compounds and in eventually reducing their toxicity. Bacterial species of different genera, including *Flavobacterium*, *Arthrobacter*, *Burkholderia*, *Plesiomonas*, *Agrobacterium*, *Pseudomonas*, *Serratia*, *Ochrobactrum*, and *Bacillus*, have been isolated with the capability to hydrolyze organophosphates (Sethunathan and Yoshida

1973; Nelson 1982; Keprasertsup et al. 2001; Zhongli et al. 2001; Harcourt et al. 2002; Liu et al. 2005; Pakala et al. 2006; Qiu et al. 2006; Yang et al. 2007).

Bacterial enzymes, such as OPH, MPH, and OpdA, encoded by opd, mpd, and opdA, respectively, are responsible for the preliminary hydrolysis of organophosphates. These enzymes perform the similar function of hydrolyzing methyl parathion into p-nitrophenol and dimethyl thiophosphate, but their amino acid sequences differ from each other. Among these bacterial enzymes, OPH (organophosphorus hydrolase) has been the most widely studied for its hydrolytic activity on organophosphates, and the gene responsible for this enzyme was given the name opd (organophosphate degrading) (Chaudhry et al. 1988; McDaniel et al. 1988; Dumas et al. 1989; Mulberry and Karns 1989; Raushel and Holden 2000). A novel gene mpd (methylparathion degrading), which encodes enzyme MPH (Methylparathion hydrolase was identified by Fu et al. (2004) and Zhongli et al. (2001). The gene mpd was found in *Plesiomonas* M6, *Pseudomonas* WBC-3, and *Stenotrophomonas* sp. YC-1 (Zhongli et al. 2001; Liu et al. 2005; Yang et al. 2006).

Another gene opdA (organophosphate degrading *Agrobacterium*) was identified in *Agrobacterium radiobacter* P230 and possessed 88.4% nucleotide sequence homology with opd gene from *Flavobacterium* sp. ATCC 27551 (Horne et al. 2002). P-nitrophenol (PNP), which is also toxic to humans, is produced from the enzymatic hydrolysis of methyl parathion, parathion, and methylparaoxon (Walker and Keasling 2002). Due to its high water solubility, PNP disperses widely in terrestrial and aquatic environments (Zaidi et al. 1996), thereby increasing the probability that it will poison animals. PNP should also be degraded along with the organophosphates. There have been reports of PNP degradation and compounds related to PNP by microorganisms (Kitagawa et al. 2004; Ningthoujam 2005; Khan et al. 2006; Qiu et al. 2009). However, to date, few bacteria have been identified that are capable of simultaneously utilizing organophosphates and PNP (Liu et al. 2005; Pakala et al. 2006; Qiu et al. 2007). The conversion and degradation of methylparathion to methyl paraoxon and p-nitrophenol are shown in Figure 10.

Table 5. Microorganisms responsible for parathion and methylparathion biodegradation

Organophosphate Pesticide	Microorganisms	References
Parathion	*Bacteria*	
	Bacillus subtilis	Yasuno et al. (1965)
	Pseudomonas melophthara	Boush and Matsumura, 1967
	Rhizobium sp	Mick and Dahm (1970)
	Flavobacterium sp ATCC 27551	Sethunathan and Yoshid (1973)
	Bacillus sp and Pseudomonas sp	Siddaramapa et al. (1973)
	Mixed bacterial culture *Pseudomonas sp* *Azotomonas sp* *Xanthomonas sp* *Brevibacterium sp*	Munnecke and Hsich (1974)
	Pseudomonas aeruginosa	Gibson and Brown (1974)
	Pseudomonas stutzeri, *Pseudomonas aeruginosa*	Daughton and Hsich (1977)
	Unidentified bacteria	Cook et al. (1978a)
	Pseudomonas sp	Rosenberg & Alexander (1979)
	Pseudomonas diminuta	Serdar et al. (1982)
	Arthrobacter sp, Bacillus sp	Nelson (1982)
	Pseudomonas sp, Xanthomonas sp	Tchelet et al. (1993)
	Agrobacterium sp	Horne et al. (2002)
	Agrobacterium sp. strain Yw12	Wang et al. (2011)
	Fungi	
	Penicillium waksmani	Rao and Sethunathan (1974)
Methylparathion	*Bacteria*	
	Bacillus subtilis	Miyamoto et al. (1966)
	Flavobacterium sp ATCC 27551	Adhya et al. (1981)
	Pseudomonas sp, *Flavobacterium sp*	Chaudhry et al. (1988)
	Bacillus sp	Sharmila et al. (1989)
	Bacillus sp	Ou and Sharma (1989)
	Unidentified bacteria	Mishra et al. (1992)
	Pseudomonas putida	Rani and Lalithakumari (1994)
	Burkholderia sp	Hayatsu et al. (2000)
	Burkholderia cepacia, Bacillus sp	Keprasertsup et al. (2001)
	Plesiomonas sp	Zhongli et al. (2001)
	Pseudomonas sp	Zhongli et al. (2002)
	Pseudomonas sp	Yali et al. (2002)
	Pseudomonas aeruginosa mpd-5	Usharani (2016) , Usharani (2021)
	Fungi	
	Trichoderma viride	Baarschers and Heitland (1986)
	Fusarium sp mpd-1	Usharani (2013), Usharani (2021)

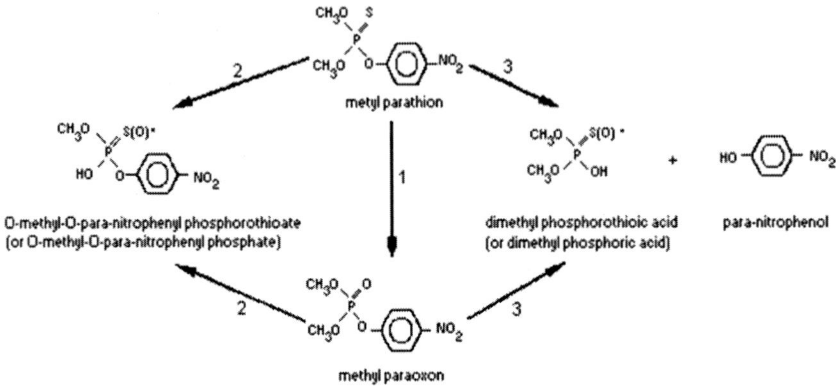

Figure 10. Methylparathion to Methylparaoxon degradation.

10.8. Organophosphate Hydrolase (OPH)

The OP trimesters and phosphofluoridates which are used as pesticides or chemical warfare agents have existed for only about fifty years. Since the early 1960s, it has been observed that soil microbial communities are capable of metabolizing pesticides by enzymatic reactions (Caldwell, 1991). OPH isolated from soil microorganisms has been shown to have a wide range of hydrolytic capabilities on OP pesticides.

Methylparathion (MP) and parathion toxicity can be reduced by nearly 120-fold by enzymatic hydrolysis, which leads to the formation of dialkylthiophosphates and PNP as byproducts (Shimazu et al. 2001). In general, the hydrolytic cleavage of the OP bond is the initial step in OP metabolism. This reaction is of considerable importance because it renders the molecule biologically inactive (Rani and Lalithakumari, 1994). OPH activity was first detected in the soil bacterial strain, *Pseudomonas diminuta,* by observation of parathion hydrolysis.

The hydrolytic enzyme activity was later found to be associated with a protein encoded by a plasmid-born gene (Caldwell, 1991). *Flavobacterium* sp. ATCC 27551 also has been shown to possess this hydrolytic enzyme activity. The *opd* gene, which expresses OPH activity, from both microbial strains was reported to be homologous by hybridization studies (Cho et al. 2000). The functional significance of the plasmid-born gene expressing OPH has not yet been reported. Nonetheless, many OP pesticides are degraded enzymatically in the soil and the resultant metabolites have been characterized. Bacteria, yeast, fungi, and mammals have shown hydrolysis of OP pesticides. However, only OPH from *Pseudomonas diminuta* has been fully characterized (Caldwell, 1991).

OPH has several advantages over many di-isopropyl-fluorophosphatases, OP hydrolases, or paraoxonases. OPH possesses much broader substrate specificity and will tolerate significant variation in the structure of its substrates. This enzyme is capable of hydrolyzing OP compounds, including those that have P-O bonds as in diethylparaoxon ("paraoxon"), P-S bonds as in malathion, P-N bonds as in acephate, and P-CN bonds as in tabun. OPH is also capable of hydrolyzing sarin, soman, and mipafox through the P-F bond (Dave et al. 1993). OPH has been shown to hydrolyze a broad spectrum of OP trimesters, thioesters, and fluorophosphonates in both whole-cell extracts and as the purified enzyme. The hydrolytic capability of this enzyme on OPs is extremely high (Dave et al. 1993). Because of the high catalytic rate of OPH and its capability to hydrolyze a wide variety of OP pesticides and chemical warfare agents, this enzyme has generated significant interest in its potential for bioremediation applications. The use of living systems where natural biological processes are responsible for degradation has been exploited in the detoxification of chemical wastes (Caldwell, 1991). This process requires the maintenance of the organism; however, the direct utilization of specific enzymes can provide another feasible alternative. The *opd* gene encoding OPH in *Pseudomonas diminuta* has been extensively studied (Hong, 1997). The gene has been expressed in *E. coli* to determine its feasibility as an active component in stable systems capable of degrading OP compounds (Hong et al. 1998).

The fungus possesses several enzymatic systems, such as glucose oxidase, catalase, lactanase (Witteveen, 1993), cytochrome P450 monooxygenase, and ligninolytic enzymes (Prenafeta boldu, 2002). Cytochrome P450 monooxygenase is coupled to NADPH reductase, which works as a source of electrons for oxidation reactions. Its system plays a central role in oxidative metabolism, as well as in the detoxification of xenobiotics. The enzyme catalyzes the epoxidation of the aromatic ring, producing arene oxides that are formed through the epoxide hydrolase trans-dihydro diols or are rearranged nonenzymatically to form phenols (Prenafeta boldu, 2002). In an oxidation study of ten organophosphorate pesticides, including methylparathion mediated by the enzyme chloroepoxidase, starting from the fungus *Caldaromyces fumago*, Hernandez et al. (1998) transformed seven of the ten pesticides into their oxone forms. They found similarities with cytochrome

P450 action but, unlike cytochrome P450, the chloroperoxidase was not capable of cleaving the oxone structures.

The commonly studied white rot fungus *Phanerochaete chrysosporium* has been shown to degrade and mineralize a wide variety of industrial and agricultural pollutants. Enzymes involved in the degradation of pollutants in *P. chrysosporium* are found to be lignin peroxidases (LiP), and manganese-dependent peroxidases (MnP), have also been shown to facilitate both reductive and lipid peroxidation-mediated degradation of environmental pollutants. It is also reported that *P. chrysosporium* is capable of degrading several chlorinated xenobiotics under conditions that do not favor the production of LiP and MnP. *P. Chrysosporium* is capable of degrading several chlorinated xenobiotics under conditions that do not favor the production of lignin peroxidizes (LiP) and manganese-dependent peroxidizes (MnP) (Kullman and Matsumura, 1996). Treatment with basidiomycetous fungi or their lignin-degrading enzymes, lignin peroxidase, manganese-dependent peroxidase, and laccases has been widely reported. Compared to most degrading enzymes of bacteria which have narrow substrate specificity, the ligninolytic enzymes of these fungi are very nonspecific and extracellular.

Consequently, white rot fungi can degrade various insoluble organic pollutants simultaneously (Han et al. 2004). *Phanerochaete chrysosporium* has emerged as a model system for studying the fungal degradation of xenobiotics (Childress et al. 1998). Cutinase is a hydrolytic enzyme that is subject to degrading cutin, the cuticular polymer (i.e., a polyester composed of hydroxy and epoxy fatty acids with usually n-C16 and n-C18) of higher plants. Unlike the other lipolytic enzymes such as lipases and esterases, cutinase can show enzymatic activity without interfacial activation. Microorganisms such as *Fusarium oxysporum* f. sp. pisi live on cutin as a sole carbon source producing extracellular cutinolytic enzymes. Several bacterial cutinases have been also isolated and characterized from a phyllosphere fluorescent *Pseudomonas putida* and *Pseudomonas mendocina*, cohabiting with a nitrogen-fixing bacteria, and Corynebacterium sp. (Lin and Kolattukudy, 1980; Dantzig et al. 1986; Sebastian et al. 1987; Murphy et al. 1996; Petersen et al. 1997). In recent years, the esterification and transesterification activities of cutinases and esterases have been largely exploited and could be applied advantageously in chemical synthesis (Gerard et al. 1993).

Chapter 11

Environmental Factors Affecting the Biodegradation Process

Pesticide degradation in soil can be influenced by both biotic and abiotic factors, which act in tandem and complement one another in the microenvironment. A bioremediation process is based on the activities of aerobic or anaerobic heterotrophic microorganisms. Microbial activity is affected by several physicochemical environmental parameters. The factors that directly impact bioremediation are energy sources (electron donors), electron acceptors, nutrients, pH, temperature, and inhibitory substrates or metabolites. Environment contaminated with various organic recalcitrant compounds is a very widespread problem throughout the world, particularly in industrialized areas. There are many reasons for organic compounds being degraded very slowly or not at all in the environment. The success of any bioremediation process depends upon the physical and chemical characteristics of the substrate, such as nutrient status and pH. And will be influenced by environmental factors such as temperature (Comeau et al. 1993) and biotic factors such as inoculum density (Ramadan et al. 1990). The optimum environmental conditions were varied for different microbial strains. Previous reports have demonstrated good agreement between studies investigating optimum conditions for biodegradation in liquid culture systems and bioremediation studies in soil or natural water matrices. This suggests that the former can provide a reliable prediction of the range of conditions in which pesticide-degrading bacteria will be active. Some of the important factors affecting the bioremediation process of pesticides are described below:

11.1. Temperature

The effect of temperature on the biodegradation of pesticides depends on the molecular structure of the pesticide. Temperature affects solubility, adsorption, and hydrolysis of pesticides in soil. The activity of soil microorganisms is stimulated by the rise in temperature (Burns, 1975 Racke et al. 1997). The lower soil or water temperature reduces the efficiency of microbial degradation of soil or water contaminants. Generally, the microbial degradation of pesticides hastens at higher soil/water temperatures. Temperature is another abiotic factor that influences the rate and extent of bioremediation since it affects microbial activity (Hong et al. 2006). It was reported that temperature might have a large role in degrading organic pollutants (Margesin and Schinner, 1997). Most *Pseudomonas* sps are unable to grow at 42°C and would be unlikely to persist at this temperature (Palleroni, 1986).

11.2. pH

The pH may affect pesticide adsorption and abiotic and biotic degradation processes. It also influences the mobility and bioavailability of pesticides in the soil (Burns, 1975 Racke et al. 1997). The rate of degradation of different pesticides varies considerably with the increase or decrease in soil or water pH. The degradation rate of various pesticides is strongly related to soil or water pH. Applying microorganisms to the environment (soil/water) for biodegradation of target pollutants, abiotic factors such as soil/water pH need to be evaluated critically, since these factors determine the survival of inoculated microorganisms. The pH condition would be significant while emergent an effective remediation strategy. Repeated application of pesticides results in the enhanced ability of the microbial population to degrade the pesticide.

11.3. Aerobic or Anaerobic Conditions

Most of the biodegradation of pesticides in aerobic conditions is fast and anaerobic conditions are slow; some compounds are not degraded anaerobically and some are degraded only partly and may give rise to toxic compounds.

11.4. Level of Nutrients and Co-Substrates

A contaminated site usually has a sub-optimal nutrient balance. The level of nutrients and the presence of co-substrates in the soil or water significantly affect the extent of microbial degradation of pesticides. Glucose was chosen because it is a primary substratum and the main carbon source for the bacteria. Glucose addition is important to improve the efficacy of bioremediation of persistent compounds like pesticides (Brajesh et al. 2004; Sampaio, 2005; Singh, 2006; Yang et al. 2008; Yugui et al. 2008), phenols (Rodrigues et al. 2007; Silva et al. 2007). Qiu et al. (2006) reported that the additional nutrients such as glucose and organic nitrogen greatly enhanced the growth of *Ochrobactrum* sp B2. Singh (2006), reports that the addition of glucose produces substances of high reactivity, which react more easily with the pollutant. Previous reports concerning the isolation of organophosphorus-degrading microorganisms suggest that the bacteria mainly degrade the compounds metabolically (Horne et al. 2002; Zhongli et al. 2001). Some reports showed that the isolated bacterium can utilize organophosphates as a source of carbon or phosphorus (Subhas and Dileep, 2003) from the hydrolysis products (Serdar and Gibson, 1985). In natural environments, the competition for carbon sources is immense and the utilization of pesticides as an energy source by this bacterium provides it with a substantial competitive advantage over other microorganisms (Malghani et al. 2009). There have been no successful cell-free studies with fungi capable of degrading organophosphorus insecticides. Among the co-substrates tested for effluent pre-treatment by fungi, glucose and sucrose were the best, when they were used at 5 to 10 gL^{1}. (Coulibaly et al. 2003). The mycelial mass yield was lower on pesticide-containing media than on glucose or sucrose-containing media. Differences in growth yield were, however not related to differences in total or relative enzyme production. (Liu et al. 2001). Marinho et al. (2011) evaluated the glucose effect on the removal of methyl parathion by *Aspergillus niger* AN400 and reported that the presence of a glucose concentration of 0.5 mgL^{-1} helped the removal of the pollutant. It is assumed that glucose can be indispensable both for the removal of MP and for cellular growth (Marinho, et al. 2011). Eventually, the fungal biomasses present many resources for the biopurification of wastewater.

11.5. Inoculum

Several researchers have demonstrated that the degradation of pesticides in soil or water can be accelerated by inoculation with appropriate microorganisms. Successful inoculation has been shown to depend on inoculum density, pollutant bio-availability, and conditions such as moisture

for the soil environment, temperature, pH, and organic matter content for both the soil and water environment (Karpouzas and Walker, 2000).
The following effective screening method of minimum inhibitory concentration test and biodegration kinetics techniques were used for selecting poterntial microorganism from the environmental samples include soil and water.

11.6. Screening and Kinetics of MP Pesticide Biodegrading Microbes (Usharani 2021)

11.6.1. Minimum Inhibitory Concentration (MIC) Test (Usharani 2021)

The screening of potential microbial strains using MIC test or plate assay. A potential bacterial strain (mpd-5) and fungal strain (mpd-1) are isolated from pesticide exposed and acclimatized agricultural soil. The original enrichment cultures were carried out in a synthetic wastewater containing mineral salts medium amended with the methylparathion as the sole source of carbon and energy. The methylparathion concentration used in this study was 0.1%, pH was adjusted using 1 N NaOH and 1 N HCl. The organisms were subsequently grown on nutrient agar medium (bacterial strain mpd-5) plates and Czapek's-Dox agar medium (fungal strain mpd1) plates to obtain single colonies. A pure monoculture of methylparathion-degrading individual strains (mpd-5 and mpd-1) are isolated by series of repeated plating on MSM with methylparathion agar plates. The Minimum inhibitory concentration (MIC) assay with plate screening method was carried out to screen methylparathion resistant bacteria using methylparathion MSM with methylparathion agar plates.
The inhibition growth zone was measured using an electronic micrometre and recorded and the inhibition percentage of bacterial and fungal growth was calculated using the below equations:

$$\%Growth = Ømp / Øb \times 100\% \text{ and } \%CI = 100\% - Growth,$$

where

Ømp represents the diameter (cm) of growth of the microorganism in the treatments exposed to methylparathion at each assessment.

Øb is the diameter of the negative control microorganism growth in each evaluation and % CI is the percentage of inhibitory growth.

V = ØC⁄t,

The growth rate of the microorganism inhibition was calculated with the use of the below equation: where "V" is the speed of growth, ØC represents the diameter (cm) of the microorganism growth (cm) and t is the incubation time (days). Based on the MIC test, the potential bacterial culture was identified based on their morphological characters and biochemical tests as given in Bergey's Manual of Determinative Bacteriology (Buchanan and Gibbons 1974; Holt et al. 1994). The methylparathion-degrading strains (mpd-5) was further isolated and identified as Pseudomonas species by Pseudomonas isolation agar medium, Ø pigment production by Kings medium and confirmation of the strain was carried out using selective medium (Centrimide Agar-CA). The identification and characterization of the isolated prominent bacterial strains were performed based on their morphological characters and biochemical tests (Buchanan and Gibbons 1974; Holt et al. 1994). And the potential fungal culture was identified based on their morphological characters using lacto phenol cotton blue staining. The identification and characterization of the isolated prominent fungal stains were performed (Clements and Shear 1957 and Booth 1977).

11.6.2. Kinetics of MP Biodegradation in MSM (Usharani, 2021)

Kinetics analysis was carried out, because the microbial cells were utilized as an enzyme system for MP (MP) biodegradation (Dykaar and Kitanidis 1996; Usharani, 2021). The Langmuir–Hinshelwood (L–H) kinetic model was used to describe the biotreatment process and biodegradation rate of MP by plotting the graph of ln (Ct/ C0) versus time, t, at different concentrations.

The simulated kinetics curves meet the first-order kinetic model was Determined using the following equation.

The obtained result are considered, where C0 is the initial concentration of MP, Ct is the concentration of MP at time t, and k is the reaction constant of the first-order reaction: First-order integrated law ln (1) (Ct⁄C0) = −kt. ln (eqn.1).

Determined using the following equation. The obtained result are considered, where C0 is the initial concentration of MP, Ct is the concentration of MP at time t, and k is the reaction constant of the first-order reaction:

First-order integrated law ln (1) (Ct/C0) = −kt. Ln (eqn.1)
ln [CMP] 0 / [CMP] t = − KMPt (eqn.2)

Rearrange this equation: ln (CMP)0/CMP)t=− KMP t [use log ln(x/y)=ln x–ln y]; ln [CMP]0–ln [CMP]t=−KMP t

ln [CMP] 0 = − KMPt + ln [CMP] t . (eqn.3)
Linear Equation: similar to straight line equation (eqn.4)

Equation of Straight Line : Y = mX + C
m=Slope; X=time=t; C=y intercept.
Compare the equations of straight line (Eq. 4) and linear equation (Eq. 3) m=Slope; i.e., m=−k=−K and X=t; A first-order reaction is an exponential decay (in terms of reactant):

[CMP] t = [CMP] 0 e –Kt (eqn.5)

The concentration of reactant [CMP] decreases exponentially over time:
Rate constant, K; unit for K = day (6) −1

ln [CMP] 0 = −KMPt + ln[CMP] t , (eqn.6)

where, C0 or [CMP]0 is the initial concentration of MP (mg/L), Ct or [CMP]t is the MP concentration after biotreatment at time, t, and k or KMP is the first-order rate constant.

11.6.3. Experimental Procedure (Usharani, 2021)

The potential bacterial and fungal strain resistant to MP was selected based on the MIC test and used for biotransformation in the biotreatment process. The efcient MP utilizer or tolerant strains (bacterial isolate mpd-5 and fungal isolate sp mpd-1) are screened for their potential in the laboratory and were used for the removal of aqueous MP by the analysis of GCMS and FTIR spectral studies. To determine the bacterial growth pattern and MP degradation efciency, the mineral salts medium amended with and without MP in presence or absence of glucose were prepared, and selected bacterial isolate was inoculated and incubated for 24 h at 30 °C on a rotary shaker at 150 rpm. The cells were removed by centrifugation (10,000 rpm for 15 min) and were transferred to sterile saline. The bacterial cell concentration of each strain was adjusted to an optical density at 600 nm (OD 600) of 0.1 and used as inoculum. Bacterial isolate mpd-5 was added as inoculum by adjusting the cell concentration to 0.1 of OD600. About 95 (mpd-5) × 108 CFUmL−1 (1 mL of 0.1OD) of the cells were used as an inoculum (bioinoculant) due to its potential stability to the concentration of MP. The optical density of the bacterial growth was found out. The rate of degradation of MP was studied in MSM-medium (250 ml in 500 ml Erlenmeyer fasks) amended with MP and without MP (1000 mgL−1). The medium was inoculated with 1 ml of bacterial isolate mpd-5. Another fask of the same composition lacking any carbon source except MP (1000 mgL−1) was inoculated to check the utilization of MP as the sole carbon and phosphorus source (Usharani 2013; Usharani and Lakshmanaperumalsamy 2016). To determine the growth pattern and MP degradation efciency and the mineral salts medium amended with and without MP in presence or absence of glucose were prepared, and the medium was inoculated with a mycelial mat on agar plugs (1 cm diameter) taken from the margins of actively growing cultures of the studied fungal strain (fungal isolate sp mpd-1) as inoculums into a series of fasks containing mineral salts medium (250 ml in 500 ml Erlenmeyer fasks) with and without MP (MP) (1000 mgL−1). Another fask of the same composition lacking any carbon source except MP (1000 mgL−1) was inoculated to check the utilization of MP as the sole carbon and phosphorus source. The fasks were incubated at 28±2 °C on a rotary shaker at 120 rpm, one fask from each series was removed at 24 h intervals up to a period of 168 h. The fungal biomass from each fask was separated by fltration method using Whatman flter paper No.1 and washed with deionized water. The dry weight of fungal biomass was determined by drying for constant weight in an oven at 50 °C in pre-weighed aluminium foil cups. The growth in terms of dry biomass was expressed in gL−1 (Usharani 2013; Usharani and Muthukumar 2013).

Chapter 12

Ozonation Process

12.1. Ozonation Process and Its Significance

Ozone (O_3) is a highly reactive gas formed by electrical discharge in the presence of oxygen (O_2). Substantial amounts of energy are required to split the stable oxygen–oxygen covalent bond to form ozone, and the ozone molecule readily reverts to elemental oxygen during the oxidation-reduction reactions. Ozone is more soluble than oxygen in water. Once ozone enters the solution, it follows two basic reaction paths: (1) direct oxidation, which is rather slow and selective, and (2) autodecomposition to the hydroxyl radical. Ozone systems are typically comprised of four basic parts: ozone generators, feed gas preparation, contacting, and off-gas disposal. In ozone generators ozone is produced from air or oxygen which is passed between two electrodes covered with a dielectric. Air or oxygen is prepared on-site or supplied at the plant. The generated ozone gas is fed into the water flow. To ensure necessary disinfection efficiency a certain predetermined contact time between ozone and water is provided in the contact chambers. Since the high concentration of ozone can be toxic the ozone off-gas escaping from the upper surfaces of water is collected and destroyed (e.g., treated with ultraviolet radiation) to neutralize the ozone to produce oxygen. A low pH favors the slow, direct oxidation reaction path involving O_3, and a high pH or a high concentration of organic matter favors the auto-decomposition route. A high concentration of bicarbonate or carbonate buffer reduces the rate of auto-decomposition by scavenging the hydroxyl radicals. This means that ozone residuals last longer at low pH and in highly buffered waters. Among the various treatment processes, ozonation looks to be the most promising method. Ozonation is of special interest because it has high oxidation potential, no oxidant residues are remaining and no increase in salt concentration occurs. Oxidation is a promising technique for removing endocrine-disrupting compounds (EDCs) from wastewater. Ozone is widely used for water and wastewater treatment for sanitation purposes, as well as for the oxidation of common pollutants. Ozone can effectively degrade some of the selected pesticides (including triazine, phosphorus ester, phenylurea herbicides, phenol carboxylic acids, and miscellaneous pesticides) when its dosage is more than the required disinfection level and the addition of hydrogen peroxide enhances the removal of all the pesticides due to the high reactivity of hydroxyl radicals (Meijers *et al.* 1995; Poche and Prados, 1995).

12.2. Addition of Ozone in Water and Wastewater Treatment

Since the discovery of ozone in 1840, this oxidant has been much used for water purification, due to its strong bactericide activity. However, a widespread application of ozone was only seen in a few countries in Europe, namely France, Germany, and Switzerland. During the last decades, ozonation has been introduced in more and more water treatment plants as a replacement for primary chlorination to control the formation of chlorination by-products. Because ozone is a strong oxidant, it can be used not only for disinfection but also for the removal (oxidation) of organic substances. Ozone is used for the treatment of water polluted with pesticides or other anthropogenic substances. Ozonation can be combined with UV and H_2O_2 treatment to increase the oxidation efficiency. Advanced oxidation processes (AOP) appear to be potentially effective methods for the removal of emerging substances like endocrine disruptors, pharmaceuticals, resin softeners, new disinfection by-products, etc. Methyl parathion is an endocrine disruptor. The US EPA (2003) reported possible endocrine disruption in mammals. The ATSDR (2001) also reported indications of weak oestrogenic activity. It has shown oestrogenic potential similar to 17*beta*-estradiol (the primary natural estrogen) in trout cells (Petit et al. 1997), and it induces the activity of aromatase, an enzyme that converts androgens to estrogen thus increasing breast cancer risk (Laville et al. 2006). Exposure to a mixture of methyl parathion and chlorpyrifos at $1/30^{th}$ the LD_{50} affected endocrine hormone levels in rats, increasing oestradiol in both males and females (Liu et al. 2006). It caused increased testosterone and decreased luteinizing hormone in the testes (Narayana et al. 2006a).

Ozone can also be added before coagulation to improve particle removal efficiency by inducing so-called microflocculation. This potential beneficial effect of preozonation is however dependent upon several factors: including water hardness and dissolved organic carbon concentration. In moderate to high dissolved organic carbon waters the coagulant dose is set by the dissolved organic carbon. Ozonation converts natural organic matter into smaller compounds, e.g., oxalic acid that adversely affects coagulation. Thus, microflocculation will not be observed for moderate to high dissolved organic carbon waters. In low dissolved organic carbon waters the coagulant dose is set by the particles and the adsorbed organic matter. Ozonation may react with adsorbed dissolved organic carbon and thereby alter the amount and conformation of adsorbed organic matter. Therefore microflocculation is most likely to occur in low dissolved organic carbon waters (Becker and OMelia, 2001). Thus, preozonation is effective in low dissolved organic carbon waters whereas it can have adverse effects in high dissolved organic carbon waters.

12.3. Ozone Chemistry

Ozone (O_3) is a highly reactive gas formed by electrical discharge in the presence of oxygen (O_2). A substantial amount of energy is required to split the stable oxygen–oxygen covalent bond to form ozone, and the ozone molecule readily reverts to elemental oxygen during the oxidatioxidation-reductionons. Ozone is more soluble than oxygen in water. Once ozone enters the solution, it follows two basic reaction paths: (1) direct oxidation, which is rather slow and selective, and (2) autodecomposition to the hydroxyl radical. A low pH favors the slow, direct oxidation reaction path involving O3, and a high pH or a high concentration of organic matter favors the auto-decomposition route. A high concentration of bicarbonate or carbonate buffer reduces the rate of auto-decomposition by scavenging the hydroxyl radicals. This means that ozone residuals last longer at low pH and in highly buffered waters. Ozone can be artificially produced by the action of high voltage discharge in air or oxygen.

$$O_2 + O \text{ ------- energy --------------> } O_3$$

Ozone is highly unstable and must be generated on-site. Its oxidation potential (-2.07V) is greater than that of hypochlorite acid (-1.49V) or chlorine (-1.36V), the latter agents being widely used in water treatment practice. Ozone is thought to decompose accordingly (Miller, 1978)

$$O_3 + H_2O \text{ ---------------> } HO_3 + OH^-$$
$$HO_3^+ + OH \text{ ---------------> } 2HO_2$$
$$O_3 + HO_2 \text{ ---------------> } HO + 2O_2$$
$$HO + HO_2 \text{ ---------------> } H_2O + O_2$$

This unstable form of oxygen breaks down into oxygen molecules and oxygen atoms which have high oxidation potential. If we examine the oxidation power of Ozone by measuring the REDOX potential will find out that O_3 is about 5 times more oxidising than oxygen & about twice as much as chlorine. These high potentials increase its reactivity with other elements and compounds. This reactivity is about 20 to 50 times more reactive than chlorine and permanganates as it is well documented in the case of the high kill rate of micro-organisms (fungus, bacteria & viruses). This high kill rate means smaller retention times, and storage tanks, are required to do the same disinfecting as other oxidants. In other words, the capital cost for building these tanks and treatment plants is reduced considerably. It will reduce chemical handling, storage, transportation, infrastructure, and production facilities. Ozone requires only electricity which is readily available from hydro, solar, wind, or fuel electric generators. In many instances, O_3 will allow decentralization of services which will provide better flexibility and better cost management.

12.4. Oxidation Mechanism

The free radicals (HO_2 and HO) react with a variety of impurities such as metal salts, organic matter including microorganisms, hydrogen, and hydroxide ions. They are more potent germicides than hypochlorite acid by factors of 10 to 100 fold and disinfect 3125 times faster than chlorine (Nobel 1980). Oxidation potential does not indicate the relative speed of oxidation nor how complete the oxidation reactions will be. Complete oxidation converts a specific organic compound to carbon dioxide and water. Oxidation reactions that take place during water treatment are rarely complete, due to the large quantity of contaminants and relatively short durations of time in which to oxidize the water pollutants. Therefore, partially oxidized organic compounds, such as aldehydes and carboxylic organic acids are produced during the relatively short reaction periods. These aldehydes and carboxylic acids can be removed by other means before complete mineralization to reduce the amount of Ozone needed for complete oxidation of these chemicals. Three fundamental mechanisms apply to the oxidation of organic compounds reacting with an oxidizer. Each mechanism is unique as to how organic compounds react with an oxidizer. But in some cases, oxidants will react with organic compounds by all three mechanisms, although in sequential steps. First is the addition mechanism which occurs with organic compounds containing aliphatic unsaturates, such as olefin. Ozone can be added across a double bond to form an ozonide. This reaction occurs readily in nonaqueous solvents, but as soon as water is added, the ozone hydrolyzes to other products, with cleavage of the former double bond. Second is the substitution mechanism involves the replacement of one atom or functional group with another. This specific reaction also can be viewed as an insertion reaction, whereby oxygen is inserted between the ring carbon and hydrogen to form the hydrogen group on the ring. Oxidation also can involve the cleavage of carbon-carbon bonds to produce fragmented organic compounds.

12.5. Ozone Systems

Ozone systems are typically comprised of four basic parts: ozone generators, feed gas preparation, contacting, and off-gas disposal. In ozone generators ozone is produced from air or oxygen which is passed between two electrodes covered with a dielectric. Air or oxygen is prepared on-site or supplied at the plant in liquid form. The generated ozone gas is fed into the water flow. To ensure necessary disinfection efficiency a certain predetermined contact time between ozone and water is provided in the contact chambers. Since the high concentration of ozone can be toxic the ozone off-gas escaping from the upper surfaces of water is collected and destroyed (e.g., treated with ultraviolet radiation) to neutralize the ozone to produce oxygen. The experimental setup of the Oozonation process (Figure 11, Usharani et al. 2010; Usharani et al. 2012 Usharani and Muthukumar, 2016). Among various water and wastewater treatment options, ozonation and ozone-based advanced oxidation processes, such as ozone/hydrogen peroxide, ozone/ultraviolet irradiation, and ozone/hydrogen peroxide/ultraviolet irradiation, are likely key technologies for degrading and detoxifying organic pollutants in water and wastewater. Ozonation is of special interest because it has high oxidation potential, no oxidant residues are remaining and no increase in salt concentration occurs (Carballa et al. 2007). Ozone-based treatments of a major group of organophosphate pesticides are reviewed. Ozonation is considered one of the most promising variations of chemical oxidation and has a long history of investigation for aqueous pesticide degradation (Reynolds et al. 1989).

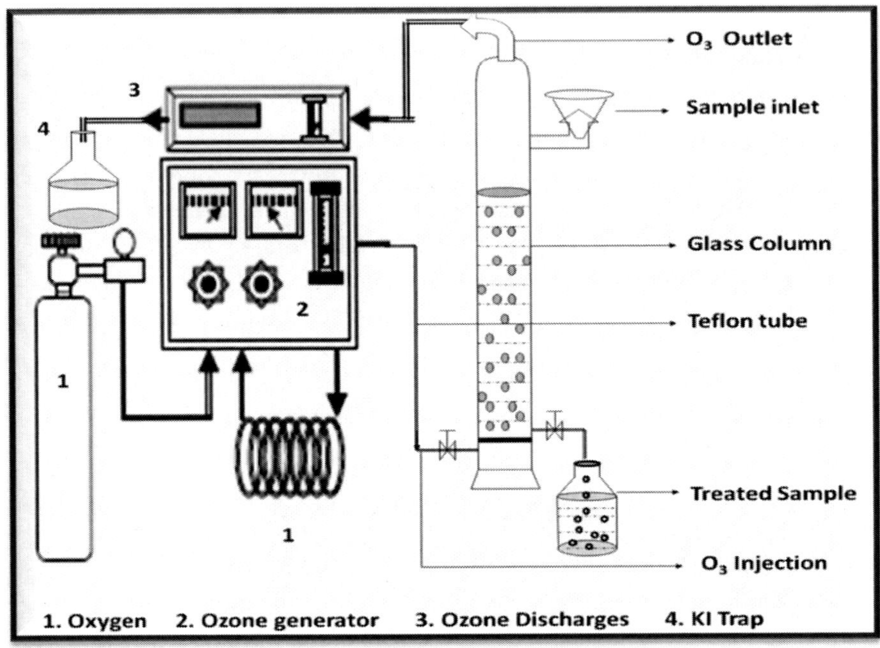

Figure 11. Experimental Set-up of the Ozonation Process (Usharani et al. 2010, 2012 and 2016).

12.6. Ozonation and Advanced Oxidation Processes (AOPs) for Wastewater Treatment

The ozonation process has been used as an effective method for removing residual pollutants such as pesticides and other hazardous chemicals from raw water during drinking water treatment. Ozone selectively reacts with compounds containing hetero atoms such as S, N, O, and Cl. Thus, pesticides, which usually have some heter atoms on the molecules, are often expected to be destroyed by ozonation. However, as has been found by many researchers, the reactivity of pesticides with ozone varies largely due to their diverse structural features (Reynolds et al. 1989; David et al. 1991).

Table 6. Advanced oxidation processes examined for aqueous pesticide degradation

Process	Oxidant(s)	Other chemical(s)	Another energy source	Note
Ozonation (pH > 8)	Ozone	Hydroxyl ion	None	
O_3/UV	Ozone	None	UV radiation	$\lambda = 258$- 260 nm for O_3
O_3/H_2O_2	O_3 , H_2O_2	None	None	
O_3/H_2O_2/UV	O_3 , H_2O_2	None	UV radiation	O_3 /UV and H_2O_2 /UV
Fenton	Hydrogen peroxide (H_2O_2)	Ferrous ion (Fe^{2+})	None	Sludge formation
Fenton-like	H_2O_2	Ferrous ion (Fe^{2+})	None	
Photo assisted Fenton (Photo Fenton)		Fe^{2+} or Fe^{3+}	Ultraviolet radiation/visible light (UV/Vis) or solar radiation	$\lambda = 400$ nm for FeIII $(OH)^{2+}$*
Photo ferrioxalate H_2O_2 (Fe^{3+} - chelate H_2O_2)	H_2O_2	Fe^{3+}, Oxalate	UV -Vis (solar) radiation	
Anodic Fenton	H_2O_2	Iron electrode	Electrical current	Fenton reactions occur only in the anode half-cell
Electrochemical Fenton (Electro Fenton)	H_2O_2	Iron electrode	Electrical current	Sludge formation
Electro photo Fenton	H_2O_2	Iron electrode	UV/Vis (solar) radiation, electrical current	
H_2O_2/UV	H_2O_2	None	UV radiation	$\lambda = 250$- 254 nm for H_2O_2*
TiO_2/hv	None	Titanium dioxide (TiO_2)	UV/Vis or solar radiation	$\lambda <400$ nm for TiO_2*
(Photolysis) †	None	None	UV radiation	
γ-Radiolysis	None	None (water)	γ – radiation	
Sonolysis	None	None (water)	Ultrasound	

*From Oppenlander (2003); † Direct photolysis is not considered AOP. However, the results are occasionally reviewed here, as they are important for comparison.

Table 7. Summary of the pesticide degradation by ozonation and ozone-based advanced oxidation processes

S.No	Name of pesticide (CAS#)	Types of process	By-products	Note: References
1	Butamifos (36335-67-8)	Ozonation	A quasi-molecular ion with 317m/z by GC-CI-MS	Oxon formation suggested (Ohashi et al. 1993)
2	Chlorfenvinphos (470-90-6)	Ozonation	N/D	
		O3/H2O2	N/D	
3	Chlorpyrifos (2921-88-2)	Ozonation	A molecular ion with 333 m/z by GC-EI-MS	Oxon formation suggested (Ohashi et al. 1994)
4	Chlorpyrifos methyl (5598-13-0)	Ozonation	A molecular ion with 333 m/z by GC-EI-MS	Oxon formation suggested (Ohashi et al. 1994)
5	Cyanofenphos (13067-93-1)	Ozonation	A molecular ion with 333 m/z by GC-EI-MS	Oxon formation suggested (Ohashi et al. 1994)
6	Cyanophos (2636-26-2)	Ozonation	A molecular ion with 333 m/z by GC-EI-MS	Oxon formation suggested (Ohashi et al. 1994)
7	Diazinon (333-41-5)	Ozonation	Degradation pathway proposed	Oxon formation suggested (Ohashi et al. 1994)
8	Dichlorvos (627-73-7)	Ozonation	Chloride (no phosphate detected)	(Kim et al. 2002)
		Ozone/microporous silicate	Chloride, Phosphate	(Kim et al. 2002)
9	Dioxabenzofos (Salithion) (3811-49-2)	Ozonation	A molecular ion with 200 m/z by GC-EI-MS	Oxon formation cleavage of methyl moiety suggested (Ohashi et al. 1994)
10	Edifenphos (17109-49-8)	Ozonation	Degradation pathway proposed	(Ohashi et al. 1994)
11	EPN (2104-64-5)	Ozonation	Oxon, phenyl phosphonic acid ethyl ester	(Ohashi et al. 1993; Ohashi et al. 1994)
12	Ethion (563-12-2)	Ozonation	Molecular ions with 352 and 368 m/z by GC-EI-MS	Oxon formation suggested (Ohashi et al. 1994)

S.No	Name of pesticide (CAS#)	Types of process	By-products	Note: References
13	trimethyl phosphate	Ozonation	Oxon, trimethyl phosphate	(Ohashi et al. 1993; Ohashi et al. 1994) Refer to Reynolds et al (1989) for earlier studies
14	Fenthion (55-38-9)	Ozonation	Degradation pathway proposed (Ohashi et al. 1994)	Refer to Reynolds et al (1989) for earlier studies
15	Glyphosate (1071-83-6)	Ozonation	N/D	
16	Isofenphos (25311-71-1)	Ozonation	A quasi-molecular ion with 330 m/z by GC-CI-MS (Ohashi et al. 1994)	Oxon formation suggested (Ohashi et al. 1994)
17	Isoxathion (18854-01-8)	Ozonation	Oxon, triethyl l phosphate	(Ohashi et al. 1993; Ohashi et al. 1994)
18	Malathion (121-75-5)	Ozonation	N/D	Refer to Reynolds et al (1989) for earlier studies
		O_3/H_2O_2	N/D	
		O_3/UV	N/D	The commercial formulation was treated (Kearney et al. 1987)
19	Methidathion (950-37-8)	Ozonation	A molecular ion with 286 m/z by GC-EI-MS	Oxon formation suggested (Ohashi et al. 1994)
20	Methylparathion (298-00-0)	Ozonation	N/D	
		O_3/H_2O_2	N/D	
		O_3/UV Ozonation, O_3	Methyl-paraoxon, p-nitrophenol (Zwiener et al. 1995) Paraoxon, P-Nitrophenol	UV irradiation was not effective (Zwiener et al. 1995) Usharani et al. 2010 and 2012; Usharani, 2023

Table 7. (Continued)

S.No	Name of pesticide (CAS#)	Types of process	By-products	Note: References
21	Mevinphos (7786-34-7)	O_3/UV	N/D	(Kuo, 2002)
		TiO_2/UV	N/D	(Kuo, 2002)
		UV Photolysis	N/D	(Kuo, 2002)
22	Parathion (956-38-2)	Ozonation	N/D	Refer to Reynolds et al (1989) for earlier studies
23	Phenthoate (Fenthoate) (2597-03-7)	Ozonation	A molecular ion with 304 m/z by GC-EI-MS (Ohashi et al. 1994)	Oxon formation suggested (Ohashi et al. 1994)
24	Phorate (298-02-2)	O3/UV	Sulfate, phosphate, carbonate, phorate sulfoxide, phorate, oxon sulfone, diethyl phosphite	(Ku and Lin, 2002)
25	Phosmet (732-11-6)	Ozonation	A molecular ion with 301 m/z by GC-EI-MS	Oxon formation suggested (Ohashi et al. 1994)
26	Pirimiphosmethyl (29232-93-7)	Ozonation	Degradation pathway proposed	(Chiron et al. 1998)
27	Tetrachlorvinphos (22248-79-9)	Ozonation	N/D	
28	Tolclofosmethyl (57018-04-9)	Ozonation	Trimethyl phosphate (Ohashi et al. 1993), a molecular ion with 284 m/z by GC-EI-MS (Ohashi et al. 1994)	Oxon formation suggested (Ohashi et al. 1994)

N/D- Not detected.

The rate constant indicates the reactivity of a reaction, and it becomes of vital importance when deciding whether ozonation is an economically sound option for removing pesticides from raw water during drinking water treatment. The rate constants for the ozonation of pesticides are usually acquired through a method of direct experimental determination. While this method provides exact results under the established experimental conditions, the results are not comparable with other rate constants obtained under different conditions. Further, direct determination of rate constants requires laborious and time-consuming techniques, which makes the field. Therefore, an alternative procedure that can be used to estimate the rate constant of a pesticide from an existing database is desirable. In addition to ozonation, various ozone-based advanced oxidation processes (AOPs), which utilize hydroxyl radicals for oxidation, were recently evaluated for the degradation of aqueous organic pollutants including pesticides (Camel and Bermond, 1998).

In this review, recent literature published in the past fifteen years concerning ozonation and ozone-based advanced oxidation treatment of several pesticides in aqueous medium is reviewed to update the information relating to the degree of reaction, reaction kinetics, identity and characteristics of oxidation by-products, and possible degradation pathways for the pesticides. Ozone has long been used for disinfection, odor management, and color removal in the water treatment industry. Molecular ozone has a higher oxidation potential of 2.07 V (relative to the hydrogen electrode) than conventional chemical oxidants such as potassium permanganate and chlorine. Ozonation is proven to be effective in degrading several recalcitrant organic pollutants in water and wastewater (Reynolds et al. 1989; Masten and Davies, 1994; Rice, 1997; Alvares et al. 2001). On the other hand, advanced oxidation processes (AOPs) are relatively new water and wastewater treatment technologies, which are characterized by the production of hydroxyl radicals through various chemical and photochemical reactions. The hydroxyl radicals produced by AOPs have a higher oxidation potential (2.8 V) than molecular ozone and can attack organic and inorganic molecules non-selectively with very high reaction rates (Schwarzer, 1995; Camel and Bermond, 1998; Andreozzi et al. 1999; Ikehata and Gamal El-Din, 2004).

The major AOPs examined for pesticide degradation are listed in Table 6. It should be noted that ozonation at a high pH (>8) is also regarded as an AOP because the decomposition of ozone molecules into hydroxyl radicals is predominant under such conditions, and the reactions between the radicals and organic molecules take place. The aqueous pesticide degradation, by ozonation and ozone-based AOPs such as O_3/H_2O_2, O_3/UV, and $O_3/H_2O_2/UV$, were reviewed in Table 7 (Ikehata and Gamal El-Din, 2005a and 2005b). In general, the O_3 molecule itself was one of the main reactive species under acidic conditions, so the degradation of MP exhibited low efficiency (Kasprzyk-Hordern et al. 2003). As the solution became more basic, the rate of O3 dissociated with secondary oxidants, such as hydroxyl radicals, (Muthukumar et al. 2004). However, the rate of degradation decreased at pH values higher than 10.0, which might be attributed to factor such that a high pH creates more free radical scavengers (i.e.CO_3^{2-}, HCO^{3-}, etc.), resulting in a decrease in the concentration of •OH (Alaton et al. 2002 and Zhao et al. 2004).

Chapter 13

Strategy of Experimentation: The Role of Response Surface Methodology (RSM)

The development of response surface methods began with the publication of a landmark article by Box and Wilson (1951) entitled "On the Experimental Attainment of Optimum Conditions." The strategy of experimentation begins with a standard two-level fractional factorial design, mathematically described as "2^{k-p}" (Box and Hunter, 1961), which provides a screening tool. During this phase, experimenters seek to discover the vital few factors that create statistically significant effects of practical importance for the goal of process improvement. To save time at this early stage where a number (k) of unknown factors must be quickly screened, the strategy calls for the use of relatively low-resolution ("Res") fractions (p). The typical tools used for RSM, are the central composite design (CCD) and Box-Behnken design (BBD). These basic concepts are the elements that allow the user to plan an effective design of experimentation (DOE) strategy. The overall strategy is identifying and/or quantifying significant variables and variable interactions. The basic approach consists of four phases: Discovery, Breakthrough, Optimization, and Validation as illustrated in Figure 12.

The appropriate phase is dependent on the researcher's goals, resources, timing, and subject matter knowledge. Note, however, that a user could sequence through all four phases within a single project. Each DOE phase offers different results. For instance, in the discovery phase, a researcher's primary goal would include 'screening' for vital input variables. However, the investigator must be aware of the inability to generate a prediction equation using low-resolution screening arrays, if any variable interactions are at work in the system. The discovery phase typically focuses on two-level, fractional-factorial arrays to identify the "vital few" variables. Fractional-factorial arrays range from resolution III to VI. The lower resolution arrays (III and IV), for instance, 2^{3-1}, and 2^{4-1} are limited in application. These arrays are used primarily to screen input variables because of their limited capability to quantify two-factor interactions. Full-factorial and resolution V and VI fractional arrays are more commonly used in the breakthrough phase. All fractional-factorial arrays are subsets of full-factorial arrays. The higher resolution fractional arrays (V and VI), such as MR-5 and 2^{6-1}, offer the opportunity to screen larger numbers of variables with a minimal number of experiments. These designs allow for the identification of some two-factor interactions with a minimal volume of additional experiments. The breakthrough phase is capable of generating first-order prediction equations which include main effects and interaction coefficients. From such data, we pursue higher-order models from response surface designs.

Subject matter knowledge and prior screening results establish a region of operability, an area where performance is achievable. In the optimization phase, you refine the selection of factors and factor ranges to aid in framing a region of interest, the area where performance peaks. In the optimization phase, variable testing often progresses to three testing levels (3^k) for significant variables. At three levels of testing, the number of experiments increases, however, coefficients for the squared component of input variables can be quantified and a higher order polynomial generated, i.e., quadratic. The higher-order equations are more accurate and allow the researcher to 'hone' the performance level for the quality characteristic(s) of interest. The ability to readily and effectively 'dial in' a desired performance can identify peak performance levels for either product or process. As indicated in the flowchart of Figure 12, response surface methods (RSM) are utilized in the optimization phase. Due to the high volume of experiments, this phase focuses on a few highly influential variables, usually 3 to 5. The typical tools used for RSM are the central composite design (CCD) and the Box-Behnken design (BBD). With the aid of software, the results of these complex designs are exhibited pictorially in 3D as 'mountains' or 'valleys' to illustrate performance peaks.

In the validation phase, we focus on confirmation of results drawn from previous phases. Typically by this point, we have identified significant variables and selected variable levels based on a targeted performance. However, before we move to a long-term application, like manufacturing, we wish to confirm our findings. In theory, we are primarily interested in main effects and significant two-factor interactions – review the 80/20 rule discussed in the May/June segment. Therefore, the focus in the validation phase is arrays exhibiting IV, V, and VI resolution levels. The exceptions, not detailed here, include complex chemical systems.

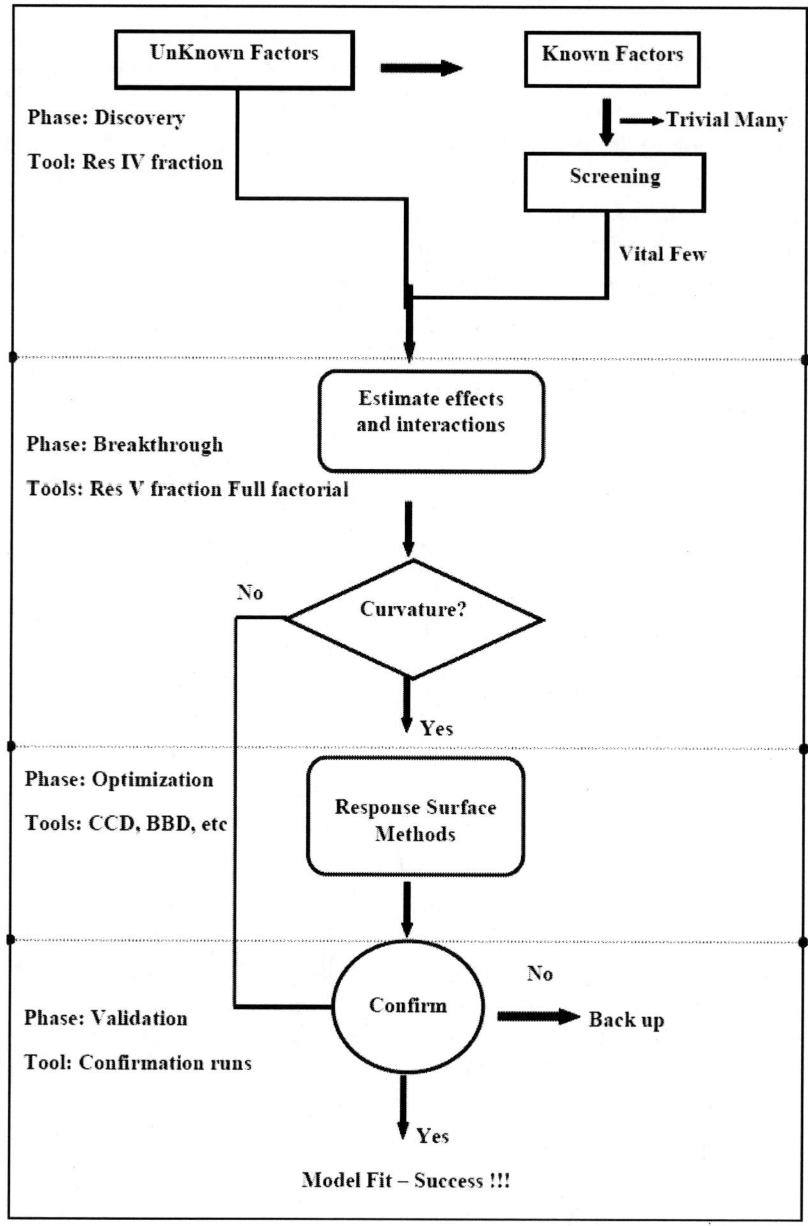

Figure 12. A flow chart of the four phases of DOE, the appropriate phase dependent on goals, resources, timing, and subject knowledge.

Selecting the proper DOE strategy is a skill typically cultivated through years of experience or learned from workshops and case studies. DOE strategies run the gambit from classic full- and fractional-factorial designs to the latest irregular factorials called minimum run (MR) arrays. These complex arrays are made practical with the availability of software that readily generates balanced designs. Understanding the risks and benefits of each strategic phase improves the user's ability to select effective designs, as well as builds their tool portfolio. DOE does require up-front planning, discipline, and a basic knowledge of statistics. However, it provides a proficiency that will arm the user with the power to maximize product and process performance and optimize resource efficiency, while simultaneously generating data paramount to minimizing product variability.

13.1. Experimental Design for Optimization of Process Variables Using Response Surface Methodology (RSM)

Response surface methodology (RSM) is a collection of statistical and mathematical techniques that uses quantitative data from appropriate experiments to determine regression model equations and operating conditions which are useful for developing, improving, and optimizing processes (Chandra et al. 2007 and Sahu et al. 2010). RSM is very useful for modeling and analyzing a complex process, allowing the construction of a model that contains a full description of the independent variables that are effective under optimal conditions (Guillou and Floros, 1993; Meilgaard et al. 2002; Meng et al. 2004). The main advantage of RSM is the small number of experimental trials needed to evaluate multiple parameters and their interactions (Chow et al. 1998), making the optimization process more efficient and cost-effective in terms of both manpower and resources. RSM has been used successfully in a variety of experimental processes (Lee et al. 2000; Kiran et al. 2007; Zhao et al. 2007; Aslan and Cebeci, 2007). Experimental designs nowadays have been regarded as one of the most favorable techniques in covering a large area of practical statistics and obtaining unambiguous results with the least expense.

Response surface methodology (RSM) helps to quantify the relationships between one or more measured responses and the vital input factors. The classical approach of changing one variable at a time to study the effects of variables on the response is a time-consuming method, particularly for multivariable systems and also when more than one response is considered. Statistical design of experiments reduces the number of experiments to be performed, considers interactions among the variables, and can be used for optimization of the operating parameters in multivariable systems. Response surface methodology (RSM) is used when only several significant factors are involved in optimization. Different types of RSM designs include 3-level factorial design, and central composite design (CCD) (Box and Wilson, 1951; Boza et al. 2000). Box-Behnken design (Singh et al. 1995) and D-optimal design (Sanchez-Lafuente et al. 2002). RSM based on Box-Behnken's design (BBD) has some advantages over other designs like central composite and full factorial design. RSM requires fewer experiment runs, and is suitable for multiple-factor experiments, searching for relationships between factors, and finding the most suitable condition and prediction of response (Seth and Chand, 2000).

A modified central composite experimental design known as the Box-Behnken design is an independent, rotatable quadratic design with no embedded factorial or fractional factorial points where the variable combinations are at the midpoints of the edges of the variable space and the center. Among all the RSM designs, the Box-Behnken design requires fewer runs than the others, e.g., 29 runs for a 4-factor experimental design. By careful design and analysis of experiments, the Box-Behnken design allows calculations of the response function at intermediate levels which were not experimentally studied, and shows the direction if one wishes to change the input levels to determine the effects on the response (Hamed and Sakr, 2001). The most popular response surface methodologies are Central Composite and Box-Behnken designs. Box-Behnken design is an efficient and creative three-level composite design for fitting second-order response surfaces. It is an independent quadratic design. The methodology is based on the construction of balanced designs that are rotatable and enable each factor level to be tested several times. Each factor or independent variable can be placed at one of three equally spaced values (coded as –1, 0, and +1). In this design, the treatment combinations are at the midpoints of the edges of the cubical design region and the center. Box-Behnken designs provide excellent predictability within the spherical design space and require fewer experiments compared to full factorial designs or central composite designs. The number of required experiments for the Box-Behnken design can be calculated according to $N = k2 + k + cp$, where k is the factor number and cp is the replicate number of the central point Box and Behnken (1960).

The application of statistical experimental design techniques in the degradation study can result in improved yields of degradation and allow the rapid and economical determination of the optimal culture conditions with fewer experiments and minimal resources. Treating each factor separately would be very time-consuming (Yang et. al. 2003). The conventional practice of one-factor optimization by maintaining constant the other factor does not provide any information for the interaction of all factors. However, there are few reports of bioremediation processes using RSM, and it is important to make use of RSM in this area to utilize manpower and resources most effectively. Through the RSM, the experimental runs were proposed. The analysis of the data after the proposed condition of the experiment can also be evaluated accordingly. The optimized condition for the highest removal of methylparathion can be obtained after the analysis of experiments.

13.2. Optimization of Methylparathion Degradation

Optimization of methylparathion degrading condition by potential microorganisms. In order to study the effect of variables on the degradation of methylparathion by biotreatment using potential bacterial strain, the process variables include pH, temperature time and agitation were optimized. Experimental design was set using the variables such as pH, temperature, time and agitation. The synthetic wastewater which consists of mineral salts medium amended with the methylparathion was set at various temperature (25-40°C), pH (5-9), time (24-168 h) and agitation (120-180 rpm) for analysis. The concentration of methylparathion used was 1000 mgL-1. The pH was adjusted using 1N NaOH and 1N HCl with the help of pH meter. During this process, estimation of various parameters such as residual methylparathion, COD removal, TOC removal and pH were analysed to measure the degradability of methylparathion. Response surface methodology (RSM) based on the Box-Behnken design of experiment was used to optimize these parameters and their interaction which significantly influenced methylparathion biodegradation.

13.3. Box–Behnken Experimental Design (BBD) of Methylparathion Bioremoval Using RSM (Usharani and Lakshmanaperumalsamy, 2016)

A four-factor, three-level Box–Behnken design was used in this study. The Box–Behnken statistical experiment design method was used to determine the effects of operating variables such as temperature, pH, time and agitation on the percentage removal of methylparathion, COD and TOC. Four important operating variables temperature (X1), pH (X2), time (X3) and agitation (X4) were considered as independent variables. The low, middle and high levels of each variable were designated as -1, 0, and +1 respectively, (Usharani and Muthukumar, 2013; Usharani and Lakshmanaperumalsamy, 2016).

A standard RSM design called Box-Behnken's Design (BBD) for biotreatmentprocess was adopted to study the influence of variables for the removal ofaqueous methylparathion. The method can reduce the number of experimental trials needed to evaluate multiple parameters and their interactions and for finding the most suitable condition and prediction of response (Box and Behnken, 1960; Myers and Montgomery, 2002).Among all the RSM designs, BBD requires fewer runs than the others, e.g., 29 runs for a 4-factor experimental design. By careful design and analysis of experiments, Box-Behnken design allows calculations of the response function at intermediate levels which were not experimentally studied and shows the direction if one wishes to change the input levels to determine the effects on the response (Hamed and Sakr, 2001, Martínez-Toledo and Rodríguez-Vázquez, 2011).

Response surface methodology (RSM) based on the BBD of experiment was used to optimize the variables and their interaction which significantly influenced methylparathion biodegradation by the individual strains of Pseudomonas aeruginosa mpd. A four-factor, three-level Box-Behnken design was used in the biotreatment process. The Box-Behnken design is an independent, rotatable quadratic design with no embedded factorial or fractional factorial points where the variable combinations are at the mid-points of the edges of the variable space and at the center. Among all statistical experiment designs, Box-Behnken design requires fewer runs than the others, e.g., 29 runs for a 4-factor experimental design. The low, middle and high levels of each variable were designated as −1, 0, and +1 respectively, as given in Table 1. For this biotreatment process, the variables and their values in brackets were three levels include temperature (25-40°C), pH (5-9), time (24-

168 h) and agitation (120-180 rpm), at constant methylparathion concentration 1000 mgL-1(0.1%). This also enabled the identification of significant effects of interactions for the batch studies. This also enabled the identification of significant effects of interactions for the batch studies. In system involving four significant independent variables X1, X2, X3, and X4, the mathematical relationship of the response of these variables can be approximated by quadratic (second degree) polynomial equation (Box and Behnken, 1960).

A total of 29 experiments were carried out. The design consists of three replicated center points, and a set of six points lying at the midpoints of each edge of the multidimensional cube (Table 2).Response functions, describing variations of dependent factors (Y) (methylparathion removal, COD removal, TOC removal, bacterial growth for bacterial with the independent variables (Xi) (temperature, pH, time and agitation) can be written as follows (Eq.1):

$$Y = b0 + \sum bixi + \sum bijxixj + \sum biix2ii... \quad (1)$$

Linear Interaction Square

where, Y is the predicted response in percentage of methylparathion removal, COD removal, TOC removal and bacterial growth in terms of optical density, bo is the offset term and bi is the linear effect while bii and bij are the square and the interaction effects, respectively. Experimental data points used in Box-Behnken statistical experiment design are presented in Table 1. The response function coefficients were determined by regression using the experimental data and the Stat-Ease Design Expert 8.0.4 program.

The response functions for percentage of methylparathion removal, COD removal, TOC removal and bacterial growth in terms of optical density were approximated by the standard quadratic polynomial equation as presented in Eq. 2.

$$Y = b0 + b1X1 + b2X2 + b3X3 + b4X4 + b11X12 + b22X22 + b33X32 + b44X42 + b12X1X2 + b13X1X3 + b14X1X4 + b23X2X3 + b24X2X4 + b34X3X4 ... \quad (2)$$

where Y is the predicted response, i.e.,

The methylparathion removal; X1, X2, X3 and X4 are the coded levels of the independent factors: temperature, pH, time and agitation. The regression coefficients are: b0 –the intercept term; b1, b2, b3and b4–the linear coefficients; b12, b13, b14, b23 b24b34 –the interaction coefficients and b11, b22, b33, b44 –the quadratic coefficients. The model evaluates the effect of each independent factor on the response. The normal practice is to test within the feasible range, so that the variation in the process does not mask the factor effect. A total of 29 trials were necessary to estimate the coefficients of the model using multiple linear regressions. Hence, about 29 treatments were conducted in the present study and analysis the variance. The data obtained from 29 experiments, were used to find out the optimum point of the process parameters using Box-Behnken Design in Response surface methodology. All the data were treated with the aid of Design Expert by Stat Ease Inc, Minneapolis (Design Expert. 8.0.4). For this bacterial biotreatment process, the methylparathion removal conditions are presented in tables according to the experimental design (Usharani and Muthukumar, 2013; Usharani and Lakshmanaperumalsamy, 2016).

13.4. Central Composite Design (CCD) of Methylparathion Removal Using RSM (Usharani and Muthukumar, 2016)

13.4.1. Optimization of Process Variables

In order to study the effect of variables such as pH and time on methylparathion degradation and COD and TOC removals during ozonation process, the pH was varied from 5 to 12 using hydrochloric acid (1N) and sodium hydroxide (1N) and ozonated for a period of 120min. The COD was estimated as per the standard procedure. The TOC was measured using TOC analyser by combustion / NDIR method. The pH of the ozone treated wastewater was adjusted and monitored using pH meter. All experiments were performed in triplicates (Usharani and Muthukumar, 2016).

13.4.2. Experimental Design of Methylparathion Removal Using Central Composite Design

RSM is a collection of statistical and mathematical techniques that uses quantitative data from appropriate experiments to determine regression model equations and operating conditions which are useful for developing, improving and optimizing processes (Box and Behnken, 1960). In this work, a standard RSM design called Central Composite Design (CCD) was applied to study the variables for the removal of aqueous methylparathion. This method can reduce the number of experimental trials needed to evaluate multiple parameters and their interactions and for finding the most suitable condition and prediction of response. Generally, the CCD consist three kinds of runs which are the 2^n factorial runs, 2^n axial runs and six center runs, where n is the number of factors. An attempt was made to optimize the degradation of the organophosphate pesticide methylparathion by ozonation using a CCD. The effects of operating variables of factors such as time and pH were investigated. The optimization response parameters, including methyl-parathion removal, COD and TOC removals and phosphate and nitrate released were determined via response surface methodology.

A two-factor, three-level Central Composite Design was used in this study. The Central Composite statistical experiment design method was used to determine the effects of operating factors such as pH and time on percent removals of COD, TOC, methylparathion. Two important operating parameters; pH (X_1) and reaction time (X_2) were considered as independent variables. The low, middle and high levels of each variable were designated as −1, 0, and +1 respectively.

The variables and their values (in brackets) were three levels, like pH (5-9) and time (40-200 min) at constant methylparathion concentration 1000 mg/L (0.1%). This also enabled the identification of significant effects of interactions for the batch studies. In system involving four significant independent variables X_1 and X_2, the mathematical relationship of the response of these variables can be approximated by quadratic (second degree) polynomial equation. This design is suitable for exploration of quadratic response surfaces and construction of a second order polynomial model, thus helping in optimizing a process using a small number of experimental runs.

A full factorial central composite design for 2 independent factors with 6 replication of the central points and 6 axial points, leading to a total of 13 sets of experiments were evaluated. Low and high factor settings were coded as –1 and +1 respectively, the centre points was coded as 0 and the design is extended up to $+\alpha$ and $-\alpha$. The value of alpha represents the distance from the centre of the design space to an axial (Usharani and Muthukumar, 2016).

Response functions describing variations of dependent variables (methylparathion, COD and TOC removals) with the independent variables (X_i) can be written as follows:

$$Y = \overbrace{\underbrace{b_o + \sum b_{i} X_i}} + \overbrace{\underbrace{\sum b_{ij} X_i X_j}} + \overbrace{\underbrace{\sum b_{ii} X_i^2}} \quad (1)$$

Linear interaction squared

Where, Y is the predicted response (percent methylparathion, COD, TOC removals), b_o is the offset term and b_i is the linear effect while b_{ii} and b_{ij} are the square and the interaction effects, respectively. Experimental data points used in Box-Behnken statistical experiment design are presented in Table 2. The response function coefficients were determined by regression using the experimental data and the Stat-Ease Design Expert 8.0.4 computer program. The response functions for percent methylparathion, COD and TOC removals were approximated by the standard quadratic polynomial equation as presented below. The following Equation describes the regression model of the present system, which includes the interaction terms:

$$\mathbf{Y = b_0 + b_1X_1 + b_2X_2 + b_{11}X_1^2 + b_{22}X_2^2 + b_{12}X_1X_2} \quad (2)$$

Where Y is the predicted response, i.e. the methylparathion removal; X_1 and X_2 are the coded levels of the independent factors: pH and time. The regression coefficients are: b_0 – the intercept term; b_1 and b_2 – the linear coefficients; b_{12} – the interaction coefficients and b_{11}, b_{22} – the quadratic coefficients. The model evaluates the effect of each independent factor on the response. The normal practice is to test within the feasible range, so that the variation in the process does not mask the factor effect (Usharani and Muthukumar, 2016).

A total of 13 trials were necessary to estimate the coefficients of the model using multiple linear regressions. Hence, about 13 treatments were conducted in the present study and analysis the variance. The data obtained from13 experiments, were used to find out the optimum point of the process parameters by using Central Composite Design in Response surface methodology [17, 18]. All the data were treated with the aid of Design Expert by Stat Ease Inc, Minneapolis (Design Expert. Stat-Ease-8.0.4) (Usharani and Muthukumar, 2016).

Conclusion

Pesticide contamination caused by either natural or anthropogenic activities is one of the most serious environmental problems. Continuous and excessive use of organophosphorus compounds has led to the contamination of several ecosystems in different parts of the world. Its widespread usage has caused environmental concern due to its frequent leakage into surface and groundwater. Industries manufacturing these pesticides release wastewater in water bodies or land. Although industries treat their wastewater by activated sludge process, no attention is paid to the removal of specific pesticides or their metabolites which exert toxicity at very low concentrations. Therefore, there is a need for economically dependable methods of organophosphate degradation and removal from the environment.

An ideal treatment method for such pesticide wastes should be non-selective, able to achieve rapid and complete degradation to inorganic compounds and be suitable for small-scale wastes. Among various methods, Ozonation is of significance because it has high oxidation potential, without leaving residues, and is widely used for wastewater treatment for decontamination purposes, as well as for the oxidation of common pollutants including pesticides. Due to environmental concerns associated with the accumulation of pesticides in water supplies and food products, there is a great need to develop safe, convenient, and economically feasible methods for pesticide remediation. For this reason, several biological techniques involving biodegradation of organic compounds by microorganisms like bacteria and fungi have been developed. The capacity of certain bacterial and fungal groups to continue to exist and grow in intense or scattered environments, such as under high levels of pesticide contamination, may be useful for hazard assessment and bioremediation of polluted environments. To date, microbial transformations have been the main focus of research on organophosphate pesticide degradation. The use of specific microorganisms adapted to these pesticides, in the treatment of industrial effluents is not in practice.

Comparing the efficiency of the treatment process revealed that the ozonation process is efficient concerning treatment time followed by bacterial and fungal biotreatment of methylparathion. By-product formation and possible complete mineralization for a high initial concentration of methylparathion were effective in the bioremediation process when compared to the ozonation process. Therefore, the coupled ozonation and biological process is a promising and economically viable technology for the treatment of wastewater containing methylparathion.

References

Abbasi, A. F., Ahmad, M. and Wasim, M. (1987), Optimization of concrete x proportioning using reduced factorial experimental technique, *ACI Mater J.,* 84(1):55–63.

Adhya, T. K., Barik, S., and Sethunathan, N. (1981), Hydrolysis of selected organophosphorus insecticides by two bacterial isolates from flooded soil, *J Appl Bacteriol.*, 50: 167–172.

Ahmad, M. A. and Alrozi, R. (2010), Optimization of preparation conditions for mangosteen peel-based activated carbons for the removal of remazol brilliant blue R using response surface methodology, *Chem. Eng. J.,* 165: 883-890.

Ahmed, M. K. and Casida, J. E. (1958), Metabolism of some organophosphorus insecticides by microorganisms, *J. Econ. Entomol,* 51: 59–63.

Alaton, I. A., Balcioglu, I. A., and Bahnemann, D. W. (2002), Advanced oxidation of a reactive dye bath effluent: comparison of O_3, H_2O_2/UV-C and TiO_2/UV-A processes, Water Res. 36 (5):1143–1154.

Alexander, M. (1994), *Biodegradation and Bioremediation,* Academic Press, San Diego, 302.

Allard A. S. and Neilson A. H. (1997), Bioremediation of Organic Waste Sites: A Critical Review of Microbiological Aspects, *Int. Biodeterior. Biodeg*, 39(4): 253-285.

Altschul, S. F., Maden, T. L., Schaffer, A. A., Zhang, J., Zhang, Z., Miller, W. and Lipman, D. J. (1997), Gapped BLAST and PSI-BLAST: a new generation of protein database search program, *Nucleic Acids Res.*, 25: 3389-3402.

Alvares, A. B. C., Diaper, C. and Parsons, S. A. (2001), Partial oxidation by ozone to remove recalcitrance from wastewaters - A review, *Environ. Technol.*, 22(4): 409–427.

American Public Health Association (APHA), (1998), *In: Standard methods for the examination of water and wastewater*, 20th ed. APHA, Washington, DC.

Anderson, P. E. and Lafuerza, A. (1992), Microbiological aspects of accelerated pesticide degradation. In: Anderson, J. P. E., Arnold, D. J., Lewis, F., Torstensson, L. (Eds.), *Proceedings of the International Symposium on Environmental Aspects of Pesticide Microbiology*, Department of Microbiology, Swedish University of Agricultural Sciences, Uppsala, 184–192.

Andreozzi, R., Caprio, V., Insola, A. and Marotta R. (1999), Advanced oxidation processes (AOP) for water purification and recovery, *Catal. Today*, 53(1): 51–59.

Ang, E. L., Zhao, H. and Obbard, J. P. (2005), Recent advances in the bioremediation of persistent organic pollutants via biomolecular engineering, *Enzyme Microb. Technol.,* 37:487-496.

Annadurai, G., Lai Yi Ling, Jiunn-Fwu Lee. (2008), Statistical optimization of medium component and growth conditions by response surface methodology to enhance phenol degradation by *Pseudomonas putida*, *J Hazard Mater.,* 151: 171–178.

Anonymous, (2000), Year Book in Pesticide Information, Food and Agricultural Organization (FAO), New Delhi, January–March Issue, 6–7. *Appl. Microbiol. Biotechnol.* 67 (5): 600–618.

Arisoy, M. (1998), Biodegradation of chlorinated organic compounds by white-rot fungi. *Bull. Environ. Contam. Toxicol*, 60: 872–876.

Aslan, N. and Cebeci, Y. (2007), Application of box-behnken design and response surface methodology for modeling of some Turkish coals, *Fuel.* 86: 90-97.

Assalin, M. R., Rosa, M. A. and Duran, N. (2004), Remediation of kraft effluent by ozonation: effect of applied ozone concentration and initial pH, *Ozone Sci. Eng.,* 26: 317–322.

ATSDR (1999), *Toxicological profile for methyl parathion.* Draft for public comment. Atlanta, GA, US.

ATSDR. (2001), *Toxicological Profile for MethylParathion, Agency for Toxic Substances and Disease Registry*, Atlanta USA.

Baarschers, W. H. and Heitland, H. S. (1986), Biodegradation of fenitrothion and fenitrooxon by the fungus *Trichoderma viride, J. Agr. Food Chem.,* 34: 707–709.

Badawy, M. I., Ghaly, M. Y. and Gad-Allah, T. A. (2006), Advanced oxidation processes for the removal of organophosphorus pesticides from wastewater, *Desalination.* 194:166-175.

Bai, Y. H., Zhou, L. and Wang, J. (2006), Organophosphorus pesticides residues in market foods in Shaanxi Area, China. *Food Chem.,* 98: 240-242.

Baker, K. H. (1994), *Bioremediation of surface and subsurface soils, in:* Baker KH, Herson DS (Eds), Bioremediation. McGraw-Hill, Inc., pp. 209-259.

Ballesteros Martin, M. M., Sanchez Perez, J. A., Acien Fernandez, F. G., Casas Lopez, J. L., Garcia Ripoll, A. M., Arques, A., Oller, I. and Malato Rodriguez, S. (2008), Combined photo-fenton and biological oxidation for pesticide degradation. Effect of photo-treated intermediates on biodegradation kinetics, *Chemosphere.* 70: 1476-1483.

Balthazor, T. M. and Hallas, L. E. (1986), Glyphosate-degrading microorganisms from industrial activated sludge, *Appl Environ Microbiol.,* 51: 432–434.

Barik, S. and Sethunathan, N. (1978), Metabolism of nitrophenols in flooded soils, *J. Environ. Qual.,* 7: 349–352.

Barik, S., (1984), Metabolism of insecticides by microorganisms, In Insecticide Microbiology, Ed. R Lal, Springer-Verlag, Berlin, 87–130.

Basha, C. A., Chithra, E. and Sripriyalakshmi, N. K. (2009), Electro-degradation and biological oxidation of non-biodegradable organic contaminants, *Chem. Eng. J.,* 149: 25-34.

Becker, W. C. and O'Melia, C. R. (2001), Ozone: its effect on coagulation and filtration, WST: *Water Supply.*, 1(4): 81-88.

Beltran, F. J. (1999), Estimation of the relative importance of free radical oxidation and direct ozonation UV radiation rates of micropollutants in water, *Ozone-Sci. Eng.,* 21: 207–228.

Beltran, F. J., Encinar, J. M. and Alonso, M. A. (1998), Nitroaromatic hydrocarbon ozonation in water, 1. Single ozonation, Ind. *Eng. Chem. Res.,* 37 (1): 25-31.

Beltran, F. J., Garcia-Araya, J. F. and Acede, B. (1994), Advanced oxidation of atrazine in water - II. Ozonation combined with ultraviolet radiation, *Water Res.,* 28: 2165–2173.

Beltran, F. J., Rivas, F. J., Gimeno, O. and Carbajo, M. (2005), Photocatalytic enhanced oxidation of fluorene in water with ozone, comparison with other chemical oxidation methods, *Int & Eng Chem Res.,* 44:3419–3425.

Bending, G. D., Friloux, M. and Walker, A. (2002), Degradation of contrasting pesticides by white rot fungi and its relationship with ligninolytic potential, *FEMS Microbiol Lett.,* 212: 59–63.

Benitez, F. J. (2003), *Ozone reaction kinetics for water and wastewater systems,* 1st. Ed. Lewis Publishers., 124.

Benitez, F. J., Beltran-Heredia, J., Acero, J. L. and Rubio, F. J. (2000), Rate constants for the reactions of ozone with chlorophenols in aqueous solutions, *J. Hazard. Mater.* B., 79 (3): 271-285.

Berchtold, S. R., Vanderloop, S. L., Suidan, M. T. and Maloney, S. W. (1995), Treatment of 2, 4-diamino toluene using a two-stage system: fluidized-bed anaerobic granular activated carbon reactor, *Wat. Environ. Res.,* 67: 1081-109.

Berger, L. R. (1988), Suicides and pesticides in Sri Lanka, *Am. J. Public Health.*, 78: 826–827.

Bhadbhade, B. J., Sarnaik, S. S. and Kanekar, P. P. (2002), Bioremediation of an industrial effluent containing monocrotophos, *Curr Microbiol.*, 45: 345–349.

Bhalerao, T. S., and Puranik, P. R. (2009), Microbial degradation of monocrotophos by *Aspergillus oryzae*, *Int Biodeterior & Biodegrad.*, 63: 503-508.

Bhatti, Z., Toda, H. and Furukawa, K. (2002), p-Nitrophenol degradation by activated sludge attached on nonwovens, *Water Res.*, 36 (5): 1135-1142.

Bhushan, B., Chauhan, A., Samanta, S. K. and Jain, R. K. (2000), Kinetics of biodegradation of p-nitrophenol by different bacteria, *Biochem Biophys Res Commun.*, 274: 626-630.

Booth, C. (1977), *Fusarium Laboratory guide to the identification of the major species. Common wealth Mycological Institute*, Kew surry, England pp: 130-153.

Bouchard, M. F., Bellinger, D. C., Wright, R. O. and Weisskopf, M. G. (2010), Attention deficit / hyperactivity disorder and urinary metabolites of organophosphate pesticides. *Pediatrics.* 125: 1216-1226.

Bourquin, A. W. (1977), Degradation of Malathion by salt-marsh microorganisms, *Appl Environ Microbiol.*, 33(2): 356–362.

Boush, B. M. and Matsumura, F. (1967), Insecticidal degradation by *Pseudomonas melaphthora*, the bacterial symbiotic of apple maggot, *J. Econ. Entomol*, 60: 918-920.

Box, G. E. P. and Hunter, J. S. (1961), The 2k−p fractional factorial designs. Part I (Corr: V5 p417), *Technometrics,* 3: 311-351.

Box, G. E. P., and Behnken, D. W. (1960), Some new three level designs for the study of quantitative variables, *Technometrics*, 2: 455–475.

Box, G. E. P., and Wilson, K. B. (1951), On the experimental attainment of optimum conditions, *J. R. Statist. Soc.,* 13: 1-45.

Boza, A., De la Cruz, Y., Jordan, G., Jauregui – Haza, U., Aleman, A. and Caraballo, I. (2000) Statistical optimization of a sustained release matrix tablet of lobenzarit disodium, *Drug Dev. Ind. Pharm.*, 26(2):1303 -1307.

Brajesh, K. S., Walker, A., Alun, W. and Denis, J. W. (2004), Biodegradation of chlorpyrifos by *Enterobacter* strain B-14 and its use in bioremediation of contaminated soils, *Appl. Environ. Microbiol*, 70(8): 4855–4863.

Bravo, R., Driskell, W. J., Whitehead, R. D., Needham, L. L. and Barr, D. B. (2002), Quantitation of dialkyl phosphate metabolites of organophosphate pesticides in human urine using GC-MS with isotopic internal standards, *J. of Anal. Toxicol,* 26 (5): 245-252.

Buchanan, R. M. and Gibbons, N. E. (1974), *Bergey's manual of determinative bacteriology.* 8th Ed. The Williams and Wilkins Company, Baltimore.

Bujacz, B., Wieczorek, P., Krzysko-Lupicka, T., Golab, Z., Lejczak, B. and Kavafarski P. (1995), Organophosphonate utilization by the wild-type strain of *Penicillium notatum, Appl Environ. Microbiol*, 61: 2905–2910.

Bumpus, J. A., Kakkar, S. N. and Coleman, R. D. (1993), Fungal degradation of organophosphorus insecticides, *Appl Biochem Biotechnol.*, 39/40: 715–726.

Burns, R. G. (1975), *Factors Affecting Pesticide Loss from Soil. In: Soil Biochemistry*, Paul, E. A. and A. D. McLaren (Eds.). Marcel Dekker, Inc., New York, USA., pp: 103-141.

Caldwell, S. R. (1991), *Mechanistic studies of phosphotriester hydrolysis by the phosphotriesterase from Pseudomonas diminuta and the detoxification of organophosphate pesticides using an immobilized phosphotriesterase.* PhD diss. College Station, Texas A&M University. Department of Biochemistry.

Caldwell, S. R., Newcomb, J. R., Schlencht, K. A. and Raushel, F. M. (1991), Limits of diffusion in the hydrolysis of substrate by the phosphotriesterase from *Pseudomonas diminuta*, *Biochemistry,* 30:7438–7444.

Camel, V. and Bermond, A. (1998), The use of ozone and associated oxidation processes in drinking water treatment, *Water Res.,* 32(11): 3208–3222.

Carballa, M., Manterola, G., Larrea, L., Ternes, T., Omil, F. and Lema, J. M. (2007), Influence of ozone pre-treatment on sludge anaerobic digestion: Removal of pharmaceutical and personal care products, *Chemosphere,* 67: 1444-1452.

Cassidy, M. B., Lee, H., Trevors, J. T., Zablotowicz, R. B. 1999. Chlorophenol and nitrophenol metabolism by *Sphingomonas* sp UG30. *J. of Ind. Microbiol. & Biotechnology*. (23):232-241.

Celik, S., Kunc, S. and Asan, T. (1995), Degradation of some pesticides in the field and effect of processing, *Analyst,* 120: 1739-1743.

Centre for Science and Environment, (2003), *Analysis of pesticide residues in soft drinks, Centre for Science and Environment*, New Delhi, pp. 13. www.cseindia.org.

Chandra, T. C., Mirna, M. M., Sudaryanto, Y. and Ismadji, S. (2007), Adsorption of basic dye onto activated carbon prepared from durian shell: Studies of adsorption equilibrium and kinetics, *Chem. Eng. J.,* 127:121-129.

Chang, Y. and Su, C. (2003), Flocculation behavior of *Sphingobium chlorophenolicum* in degrading pentachlorophenol at different life stages, *Biotechnol. Bioeng.* 82 (7): 843-850.

Chanika E., Soueref E., Georgiadou D, Karas P., Karanasios E., Tsiropoulos N., Tzortzakakis E. and Karpouzas D. G. (2010), Isolation of soil bacteria able to hydrolyze both organophosphate and carbamate pesticides, *Bioresour Technol.,* 102 (3): 3184-3192.

Charles, R. H. and Kennneth, V. T. (1999), *Fundamentals concepts in the design of experiments*, University press, Oxford.

Chatonnet, A., Hotelier, T. and Cousin, X. (1999), Kinetic parameters of cholinesterase interactions with organophosphates: retrieval and comparison tools available through ESTHER database: ESTerases, alpha/beta Hydrolase Enzymes and Relatives, *Chem Biol Interact.*, 14; 119-120:567-76.

Chaudhry, G. R., Ali, A. N. and Wheeler, W. B. (1988), Isolation of a methylparathion degrading *Pseudomonas* sp that Possess DNA Homologous to the opd Gene from a *Flavobacterium* sp, *Appl Environ Microbiol.*, 54, 2: 288–293.

Chauhan, A., Chakraborti, A. K. and Jain, R. K. (2000), Plasmid encoded degradation of p-nitrophenol and 4-nitrocatechol by *Arthrobacter protophormiae*, *Biochem Biophy Res Commun.,* 270(3): 733–740.

Chelme-Ayala, P., Gamal El-Din, M. and Smith, D. W. (2010), Kinetics and mechanism of the degradation of two pesticides in aqueous solutions by ozonation, *Chemosphere,* 78: 557-562.

Chen, T. F., Doong R. and Lei, W. G. (1998), Photocatalytic degradation of parathion in aqueous TiO_2 dispersion: the effect of hydrogen peroxide and light intensity, *Wat. Sci. Technol.,* 37 (8): 187–194.

Chen, W. R., Wu, C., Elovitz, M. S., Linden, K. G. and (Mel) Suffet, I. H. (2008), Reactions of thiocarbamate, triazine and urea herbicides, RDX and benzenes on EPA Contaminant Candidate List with ozone and with hydroxyl radicals, *Water Res.*, 42:137–144.

Childress A. M., Bennett J. W., Connick W. J. and Daigle D. J. (1998), Formulation of filamentous fungi for bioremediation, *Biotechnol. Techniq,* 12(3): 211–214.

Chiron, S., Fernandez Alba A. R. and Rodriguez, A. (1997), *Trends in analytical chemistry*, Vol. 16. No. 9, Elsevier Sciences. PIIS0165-9936 (97) 00078-2.

Chiron, S., Fernandez Alba A. R., Rodriguez, A. and Garcia-Calvo, E. (2000), Pesticide chemical oxidaion: State of Art, *Wat. Res.,* 34: 366-377.

Chiron, S., Rodriguez, A. and Fernandez-Alba, A., (1998), Application of gas and liquid chromatography mass spectrometry to the evaluation of pirimiphos methyl degradation products in industrial water under ozone treatment, *J. Chromatogr. A.,* 823(1–2): 97–107.

Cho, T. H., Wild, J. B. and Donnelly, K. C. (2000), Utility of organophosphorus hydrolase for the remediation of mutagenicity of methyl parathion, *Environ. Toxicol. Chem.,* 19:2022-2028.

Cho, T. M. (2001), *Bioremediation of organophosphorus neurotoxicity and genotoxicty in vitro by organophosphorus hydrolase*, PhD diss., College Station, Texas A&M University.

Chow, E. T. S., Wei, L. S., DeVor, R. E. and Steinberg, M. P. (1998), Performance of ingredients in a soybean whipped topping, *J. Food Sci.,* 52: 1761-1765.

Chu, W. and Wong, C. C. (2003), A disappearance model for the prediction of trichlorophenol ozonation. Chemosphere, 51(4): 289-294.

Chu, W., Chan, K. H. and Graham, N. J. D. (2006), Enhancement of ozone oxidation and its associated processes in the presence of surfactant: Degradation of atrazine, *Chemosphere.*, 64: 931-936.

Chu, W. and Ching, M. H. (2003), Modeling the ozonation of 2,4-dichlorophenoxyacetic acid through a kinetic approach, *Water Res.,* 3: 39–46.

Cisar, J. L. and Snyder, G. H. (2000), *Fate and management of turfgrass chemicals*, ACS Symp, Series., 743:106–126.

Clements, F. E. and Shear, J. L. (1957), *The Genera of Fungi*, Honfer Publishing, Co., New York.

Comeau, Y., Greer, C. W. and Samson, R. (1993), Role of inoculum preparation and density on the bioremediation of 2,4-D contaminated soil by bioaugmentation, *Appl. Microbiol. Biotechnol,* 38: 681–687.

Cook, A. M, Daughton, C. G. and Alexander, M. (1978a), Phosphonate utilization by bacteria, *J Bacteriol.,* 133: 85–90.

Cook, A. M., Alexander, M. and Daughton, C. G. (1980), Desulfuration of dialkyl thiophosphoric acids by a *Pseudomonad*, *Appl Environ Microbiol.*, 39: 463–465.

Cook, A. M., Daughton, C. G., and Alexander, M. (1978b), Phosphorus containing pesticide breakdown products: quantitative utilization as phosphorus source for bacteria, *Appl Environ Microbiol.,* 36: 668–672.

Cork, D. J. and Krueger, J. P. (1991), Microbial degradation of herbicides and pesticides, *Adv. Appl. Microbiol.*, 36: 1– 66.

Coulibaly, L., Gourene, G. and Agathos, N. S. (2003), Utilization of fungi for biotreatment of raw wastewaters, *African J. Biotecnol.,* 2 (12): 620-630.

Council directive 98/ 83/ EC of 3 November (1998) on the quality of water intended for human consumption, *Official Journal of the European Communities*, L 330/32.

Crawford R. L. and Ederer, M. M. 1999. Phylogeny of *Sphingomonas* species that degrade pentachlorophenol. *J. of Ind. Microbiol. and Biotechnology*. 23(320-325).

Crawford, R. L. (1987), *Biodegradation of pentachlorophenol.* U.S. Patent No. 4713340.

Daisley, H. and Hutchinson, G., 1998. *Paraquat poisoning*, Lancet, 352: 1393–1394.

Dalal, M., Dani, R. G. and Kumar, P. A. (2006), Current trends in the genetic engineering of vegetables crops, *Sci. Hortic.*, 107: 215-225.

Dantzig, A. H., Zuckerman, S. H. and Andonov-Roland, M. M. (1986), Isolation of a *Fusarium* mutant reduced in cutinase activity and virulence, *J Bacteriol.*, 168: 911-916.

Daughton, C. G. and Hsieh, D. P. (1977), Parathion utilization by bacterial symbionts in a chemostat, *Appl. Environ. Microbiol*, 34: 175–184.

Dave, K. I., Lauriano, C., Xu, B., Wild, J. R. and Kenerley, C. M. (1994), Expression of organophosphate hydrolase in the filamentous fungus *Gilocladium virens*, *Appl. Microbiol. Biotechnol,* 41: 352-358.

David C. C. Yao and Haag W. R. (1991), Rate constants for direct reactions of ozone with several drinking water contaminants, *Wat. Res.,* 25(7): 751-773.

Davis, R. F., Johnson, A. W. and Wauchope, R. D. (1993), Accelerated degradation of fenamiphos and its metabolites in soil previously treated with fenamiphos, *J Nematol.,* 25: 679–685.

De La Rochebrochard d'Auzay, S., Brosillon, S., Fourcade, F. and Amrane, A. (2007), Integrated process for degradation of amitrole in wastewaters: photocatalysis/biodegradation, *Int. J. Chem. React. Eng.,* 5, A51.

Dempsey, D. A., Silva, H. and Klessig, D. F. (1998), Engineering disease and pest resistance in plants, *Trends Microbiol.,* 285, 375–379.

Department of Health and Human Services, Public Health Service, Agency for Toxic Substances and Disease Registry.

Deshpande, N. M., Dhakephalkar, P. K. and Kanekar, P. P. (2001), Plasmid-mediated dimethoate degradation in *Pseudomonas aeruginosa* MCMB-427, *Lett Appl Microbiol.*, 33: 275–279.

DeWaard, M. A. (1974), Mechanism of action of the organophosphorus fungicide pyrazophos, Meded Landb, *Wageningen*, 74: 14–17.

Dick, R. E. and Quinn, J. P. (1995), Glyphosate-degrading isolates from environmental samples: occurrence and pathway of degradation, *Appl Microbiol Biotechnol.,* 43: 545–550.

Douglas, M. M. and Dennis, P. H. H. (1974), Microbial decontamination of parathion and p-nitrophenol in aqueous media, *Appl. Microbiol.*, 28: 212-217.

Doung, M., Penrod, S. and Grant, S. (1997), Kinetics of p-nitrophenol degradation by *Pseudomonas* sp.: an experiment illustrating bioremediation, *J. Chem. Educ.*, 74 (12): 1451-1454.

Dumas, D. P., Caldwell, S. R., Wild, J. R. and Raushel, F. M. (1989), Purification and properties of phosphotriesterase from *Pseudomonas diminuta*, *J. Biol. Chem.,* 264: 19659-19665.

Dumas, D. P., Durst, H. D., Landis, W. G., Raushel, F. M. and Wild, J. R. (1990), Inactivation of organophosphorus nerve agents by the phosphotriesterase from *Pseudomonas diminuta. Arch. Biochem. Biophys*, 277:155-159.

Eddleston, M. (2000), Patterns and problems of deliberate self-poisoning in the developed world, *Q. J. Med.*, 93: 715–731.

Elliot, A. J. and Mccracken, D. R. (1989), Effect of temperature on $O^{\cdot-}$ reactions and equilibria: a pulse radiolysis study, *Radiat Phys Chem.*, 33:69–74.

Environmental Protection Agency (EPA), (1979), National study of hospital admitted pesticide poisonings for 1974 to 1976. Washington DC, US Government Printing Office.

Errampalli, D., Tresse, O., Lee, H. and Trevors, J. T. (1999), Bacterial survival and mineralization of p-nitrophenol in soil by green fluorescent protein-marked *Moraxella* sp G21 encapsulated cells, *FEMS Microbiol Ecol.*, 30: 229-236.

Eskenazi, B., Bradman, A. and Castorina, R., (1999), Exposures of children to organophosphate pesticide and their potential adverse health effects, *Environ. Health Perspect,* 107(3):409–419.

Evans, F. L. (1972), *Ozone in Water and Wastewater Treatment,* Ann Arbor, MI: Ann

Evans, M. (2003), *Optimisation of Manufacturing Processes: A response surface approach*, Carlton House Terrace, London SW1Y 5DB. Maney Publishing, London.

Extoxnet. (1996), Pesticide information profiles, methyl parathion, Extension Toxicology Network, Available at http://extoxnet.orst.edu/pips/methylpa.htm.

Fang, H., Xiang, Y. Q. and Hao, Y. J. (2008), Fungal degradation of chlorpyrifos by *Verticillium* sp. DSP in pure cultures and its use in bioremediation of contaminated soil and pakchoi, *In Biodeterior Biodeg*., 61: 294–303.

FAO. (2000), *Pesticide Disposal Series, Baseline study on the problem of obsolete pesticide stocks,* Rome.

FAO/WHO (1996) *Pesticide residues in food — 1995 evaluations. Part II — Toxicological and environmental*. Geneva, World Health Organization, Joint FAO/WHO Meeting on Pesticide Residues (WHO/PCS/96.48).

Felsot, A. S. (1989), Enhanced biodegradation of insecticides in soil: implications for agrosystems, *Annu. Rev. Entomol*., 34: 453–476.

Feng, X., Ou, L. T. and Ogram, A. (1997), Plasmid-mediated mineralization of carbofuran by Sphingomonas sp. strain CF06, *Appl Environ Microbiol*, 63:1332-1337.

Ferris, I. G. and Lichtenstein, E. P. (1980), Interactions between agricultural chemicals and soil microflora and their effects on the degradation of ^{14}C parathion in a cranberry soil, *J. Agric. Fd. Chem.,* 28:1011-1019.

Foster, L. J. R., Kwan, B. H. and Vancov, T. (2004). Microbial degradation of the organophosphate pesticide, Ethion, *FEMS Microbiol. Lett,* 240: 49–53.

Fu, G., Cui, Z., Huang, T. and Li, S. (2004), Expression, purification, and characterization of a novel methyl parathion hydrolase, *Protein Expr Purif*., 36:170–176.

Furniss, B. S., Hannford, A. J., Smith, P. W. G. and Tatchell, A. R. (1992), *Vogel's Textbook of Practical Organic Chemistry*, fifth ed. ELBS and Longman, London.

Gan, J. and Koskinen, W. C. (1998), Pesticide fate and behaviour in soil at elevated concentrations. In P. C. Kearney and T. Roberts (ed.), *Pesticide Remediation in Soils and Water*, John Wiley & Sons, Chichester, England. pp. 59-84.

Garcia, S. J., Abu-quare, A. W, O'Connell, M, Borton, A. J. and Abou-Donia, M. B. (2003), Methyl parathion: a review of health effects, *J. Toxicol. Environ. Health,* part B. 6:185-210.

Gauger, W. K., MacDonald, J. M., Adrian, N. R., Matthees, D. P. and Walgenbach, D. D. (1986), Characterization of a streptomycete growing on organophosphate and carbamate insecticides, *Arch. Environ Contam. Toxicol,* 15: 137–141.

Gaur, M. S., Sharma, A. K., Sharma, P., Mishra, V. and Tiwari, R. K. (2008), Spectrophotometric assessments of methylparathion in water samples, Can *J Pure & Appl Sci.,* 2(3): 581-587. SENRA Academic Publishers, Burnaby, British Columbia.

Gerard, H. C., Fett, W. F., Osman, S. F. and Moreau, R. A. (1993), Evaluation of cutinase activity of various industrial lipases, *Biotechnol Appl Biochem*., 17: 181- 189.

Ghanem, K. M, Al-Garni, S. M, Al-Shehri, A. N. (2009), Statistical optimization of cultural conditions by response surface methodology for phenol degradation by a novel *Aspergillus flavus* isolate, *Afr J Biotechnol*., 8(15): 3576–3583.

Gibson, W. L. and Brown, L. R. (1974), The metabolism of parathion by *Pseudomonas aeruginosa*, Dev. *Ind. Microbiol*., 16:17-87.

Gill, I. and Ballesteros, A. (2000), Degradation of organophosphorus nerve agents by enzyme-polymer nanocomposites: efficient biocatalytic materials for personal protection and large-scale detoxication, *Biotechnol Bioeng.,* 70: 400-410.

Gilliom, R. J. B., Barbash, J. E., Kolpin, D. W. and Larson, S. J. (1999), Testing water quality for pesticide pollution, *Environ. Sci. Technol.*, 33: 164A-169A.

Gong, R., Li, N., Cai, W., Liu, Y. and Jiang, J. (2010), α-Ketoglutaric acid-modified chitosan resin as sorbent for enhancing methylene blue removal from aqueous solutions, *Int. J. Environ. Res.*, 4 (1): 27-32.

Guha, A., Kumari, B., Bora, T. C. and Roy, M. K. (1997), Possible involvement of plasmid in degradation of Malathion and chlorpyrifos by *Micrococcus* sp, *Folia Microbiol.,* 42: 574–576.

Guillou, A. A. and Floros, J. D. (1993), Multiresponse optimization minimizes salt in natural cucumber fermentation and storage, *J Food Sci.,* 58: 1381-1389.

Gunnell, D., Eddleston, M., Phillips, M. R. and Konradsen, F. (2007), The global distribution of fatal pesticide self-poisoning: systematic review, *BMC Public Health.,* 7: 357-395.

Gunner, H. B. and Zuckerman, B. M. (1968), Degradation of diazinon by synergistic microbial action, *Nature,* 217: 1183-1184.

Gupta, P. K. (2004), *Pesticide exposure -Indian scene- toxicology*, vol. 198, (1-3): 83-90.

Haddad, L. and Winchester, J. (1983), *Clinical management of poisoning and overdose. Philedelphia,* WB Saunders, 575-586.

Hamed, E. and Sakr, A. (2001), Application of multiple response optimization technique to extended release formulations design, *J. Control. Release* 73:329–338.

Han M. J., Choi H. T. and Song H. G. (2004), Degradation of Phenanthrene by *Trametes versicolor* and Its Laccase, *The Journal of Micro.,* 42(2): 94-98.

Hanrahan, G. and Lu, K. (2006), Application of factorial and response surface methodology in modern experimental design and optimization, *Crit. Rev. Anal. Chem.,* 36: 141–151.

Harcourt, R. L., Horne, I., Sutherland, T. D., Hammock, B. D. and Russell, R. J. (2002), Development of a simple and sensitive fluorimetric method for isolation of coumaphos-hydrolysing bacteria, *Lett Appl Microbiol.*, 34:263–268.

Harper, L. L., Mcdaniel, C. S., Miller, C. E. and Wild, J. R. (1988), Dissimilar plasmids isolated from *Pseudomonas diminuta* MG and a *Flavobacterium* sp. (ATCC 27551) contain identical opd genes, *Appl. Environ. Microbiol*, 54:2586–2589.

Harris, C. A., Henttu, P., Parker, M. G. and Sumpter, J. P. (1997), The estrogenic activity of phthalate esters *in vitro, Environ. Health Perspect,* 105: 802-808.

Hasan, H. A. (1999), Fungal utilization of organophosphate pesticides and their degradation by *Aspergillus flavus* and *A. sydowii* in soil, *Folia Microbiologiya.,* 44: 77–84.

Hashmi, I., Khan, M. A. and Jong, G. K. (2004), Malathion degradation by *Pseudomonas* using activated sludge treatment system (Biostimulator), *Biotechnol,* 3: 82-89.

Hayatsu, M., Hirano, M. and Tokuda, S. (2000), Involvement of two plasmids in fenitrothion degradation by *Burkholderia* sp. strain NF1000, *Appl Environ Microbiol.*, 66: 1737–1740.

Hayes, W. J. and E. R. Laws (Ed.). (1990), *Handbook of Pesticide Toxicology, Vol. 3, Classes of Pesticides.* Academic Press, Inc., NY.

He, H. Y. (2008), Photo-catalytic degradation of methyl orange in water on CuS-Cu2S powders, *Int. J. Environ. Res.,* 2 (1): 23-26.

Hernandez, J., Norma R. Robledo, Luis Velasco, Rodolfo Quintero, Michael A. Pickard, Rafael Vazquez-Duhalt (1998), Chloroperoxidase-mediated oxidation of organophosphorus pesticides. *Pestic. Biochem. Physiol.,* 61: 87-94.

Hertel, R. (1993), *Environmental health criteria 145, methyl parathion. United Nations Environment Programme*, International Labour Organisation and the World Health Organization. http://www.inchem.org/documents/ehc/ehc/ehc145.htm. Accessed on 19 Aug. 1991.

Holt, J. G., Krieg, N. R., Sneath, P. H. A., Staley, J. T. and Williams, S. T. (2000). *Bergey's manual of determinative bacteriology*, 9th. Edn. Williams & Wilkins, Baltimore, USA.

Holt, J. G., Krieg, N. R., Sneath, P. H. A., Staley, J. T., and Williams, S. T. (1994), *Bergey's manual of determinative bacteriology*, 9th. Edn. Williams & Wilkins, Baltimore, USA.

Hong, L. S., Ling, L. H., Xing, H. G., Jian, H. H., Jian, S. X. and Hong, W. Y. (2006), Respiratory substrate availability plays a crucial role in the response of soil respiration to environmental factors, *Appl. Soil Ecol.*, 32: 284–292.

Hong, M. S. (1997), *Bioremediation and neurotoxicological characterization of organophosphorus compounds*. PhD diss. College Station, Texas A&M University. Department of Chemical Engineering.

Hong, M. S., Rainina, E., Grimsley, J., Dale, B. and Wild, J. (1998), Neurotoxic organophosphate degradation with polyvinyl alcohol gel-immobilized microbial cells, *Bioremediation J.*, 2:145-512.

Hong, P. K. A. and Zeng, Y. (2002), Degradation of pentachlorophenol by ozonation and biodegradability of intermediates, Water Res., 36 (17): 4243-4254. http://www.pic.int/incs/crc1/s19add4.

Hong, Q., Zhang, Z. H., Hong, Y. F. and Li, S. P. (2007), A microcosm study on bioremediation of fenitrothion-contaminated soil using *Burkholderia* sp FDS-1, *Int Biodeter Biodeg.*, 59:55–61.

Hong, S. B. and Raushel, F. M. (1996), Metal-substrate interactions facilitate the catalytic activity of the bacterial phosphotriesterase, *Biochemistry,* 35: 10904-10912.

Horne, I., Sutherland, T. D., Harcourt, R. L., Russell, R. J. and Oakeshot, J. G. (2002), Identification of an opd (organophosphate degradation) gene in an *Agrobacterium* isolate, *Appl Environ Microbiol.*, 68:3371–3376.

Huang, W. H., Fang, G. G., and Wang, C. C. (2005), A nanometer-ZnO catalyst to enhance the ozonation of 2,4,6-trichlorophenol in water, Colloids Surf. A., 260 (1): 45–51.

Hurst, C. J., Knudsen, G. R., McInerney, M. J., Stetzenbach, L. D. and Walter, M. V. (1997), *Manual of Environmental Microbiology*. Washington, D.C., American Society for Microbiology.

Ikehata, K. and Gamal El-Din, M. (2004), Degradation of recalcitrant surfactants in wastewater by ozonation and advanced oxidation processes: A review, *Ozone Sci. Eng.,* 26(4):327–343.

Ikehata, K. and Gamal El-Din, M. (2005a), Aqueous pesticide degradation by ozonation and ozone-based advanced oxidation processes: a review (Part I), *Ozone-Sci. Eng.,* 27(2): 83–114.

Ikehata, K. and Gamal El-Din, M. (2005b), Aqueous pesticide degradation by ozonation and ozone-based advanced oxidation processes: a review (Part II), *Ozone-Sci. Eng.*, 27(3): 173–202.

Imran, H., Altaf, K. M. and Kim, J. G. (2004), Malathion degradation by *Pseudomonas* using activated sludge treatment system (bio stimulator), *Biotechnol.,* 3,:82–89.

IPCS (1992) *Methyl parathion.* Geneva, World Health Organization, International Programme on Chemical Safety (Environmental Health Criteria 145).

Jacob, G. S., Kimack, N. M., Kishore, G. M., Halllas, L. E., Garbow, J. R. and Schaefer, J. (1988), Metabolism of glyphosate in *Pseudomonas* sp. strain Lbr, *Appl Environ. Microbiol,* 54: 2953–2958.

Jain, R. K., Dreisbach, J. H. and Spain, J. C. (1994), Biodegradation of p-nitrophenol via 1,2,4-benzenetriol by an *Arthrobacter* sp, *Appl. Environ. Microbiol,* 60:3030-3032.

Jiang, J., Zhang, R., Li, R., Gu, J. D., and Li, S. (2007), Simultaneous biodegradation of methylparathion and carbofuran by a genetically engineered microorganism constructed by mini-Tn5 transposon. *Biodegradation* 18:403–412.

Jilani, S. and Altaf Khan, M. (2006), Biodegradation of cypermethrin by *Pseudomonas* in a batch activated sludge process, *Int. J. Environ. Sci. Tech.,* 3 (4): 371-380.

Johnson, J. L., Gerhardt, P., Murray, R. E. G., Wood, W. A., Krieg, N. R. (1994), Similarity analysis of rRNA Methods for general and molecular bacteriology. *American Society for Microbiology*, 68: 683-700.

Joseph, A., Abraham, S., Muliyil, J. P., George, K., Prasad, J., Minz, S., Abraham, V.J. and Jacob, K. S. (2003), Evaluation of suicide rates in rural India using verbal autopsies, 2994-9, *BMJ,* 24:1121–1122.

Kadiyala, V. and Spain, J. C. (1998), A two-component monooxygenase catalyze both the hydroxylation of p-nitrophenol and the oxidative release of nitrite from 4-nitrocatechol in *Bacillus sphaericus* JS905, *Appl. Environ. Microbiol*, 64: 2479–2484.

Kaloyanova, S. and Tarkowski, S. (1981), *Toxicology of Pesticides*, 1st ed., WHO: Copenhagen.

Kanaly, R. A., Kim, I. S., and Hur, H. G. (2005), Biotransformation of 3-methyl-4-nitrophenol, a main product of the insecticide fenitrothion, by *Aspergillus niger*, *J Agr and Food Chem.*, 53, 6426–6431.

Kanekar, P. P., Bhadbhade, B. J., Deshpande, N. M. and Sarnaik, S. S. (2004), Biodegradation of organophosphorus pesticides, *Proct Indian natn Sci Acad B.*, 70 (1): 57-70.

Kaneva, I., Mulchandani, A. and Chen, W. (1998), Factors influencing parathion degradation by recombinant *Escherichia coli* with surface-expressed organophosphorus hydrolase, *Biotechnol. Prog,* 14: 275–278.

Karalliedde, L. and Senanayake, N. (1999), Organophosphorus insecticide poisoning, *J Int Fed Clin Chem.,* 11: 4-9.

Karas, P., Perucchon, C., Exarhou, C., Ehaliotis, C. and Karpouzas D. G. (2011), Potential for bioremediation of agro-industrial effluents with high loads of pesticides by selected fungi. *Biodegradation*, 22: 215-228.

Karns, J. S., Muldoon, M. T., Mulbry, K. K., Derbysihire, K. K. and Kearney, P. C. (1987), Use of microorganisms and microbial systems in the degradation of pesticide, *ACS Symp. Ser.*, 334: 156–176.

Karpouzas, D. G. and Walker, A. (2000), Aspects of the enhanced biodegradation and metabolism of ethoprophos in soil, *Pest Manag Sci.,* 56: 540-548

Karpouzas, D. G. and Walker, A. (2000), Factors influencing the ability of *Pseudomonas putida* strains epI and II to degrade the organophosphate ethoprophos, *J Appl. Microbiol.,* 89: 40-48

Karpouzas, D. G., Hatziapostolou, P., Papadopoulou-Mourkidou, E., Giannakou, I. O.and Georgiadou, A. (2004a). The enhanced biodegradation of fenamiphos in soils from previously treated sites and the effect of soil fumigants, *Environ, Toxicol, Chem.,* 23: 2099–2107.

Karpouzas, D. G., Karanasios, E. and Menkissoglu-Spiroudi, U. (2004b), Enhanced microbial degradation of cadusafos in soils from potato monoculture: demonstration and characterization, *Chemosphere,* 56: 549-559.

Karpouzas, D. G., Morgan, J. A. W. and Walker, A. (2000), Isolation and characterization of 23 carbofuran degrading from soils from distant geographical areas, *Lett. Appl. Microbiol*., 31(5): 353-358.

Karpouzas, D. G., Morgan, J. A. W. and Walker, A. (2000), Isolation and characterization of ethoprophos-degrading bacteria, *FEMS Microbiol Ecol.,* 33: 209-218.

Karpouzas, D., Fotopoulou, A., Menkissoglu-Spiroudi, U. and Singh, B. K. (2005), Non-specific biodegradation of the organophosphorus pesticides cadusafos and ethoprophos by two bacterial isolates, *FEMS Microbiol Ecol.,* 53: 369–378.

Karpouzas, D. G. and Singh, B. K. (2006), Microbial degradation of organophosphorus xenobiotics: metabolic pathways and molecular basis, *Adv. Microb Physiol.*, 51: 119-186.

Karpouzas, D. G., and Walker, A. (2000), Factors influencing the ability of *Pseudomonas putida* epI to degrade ethoprophos in soil, *Soil Biol. & Biochem.,* 32: 1753-1762.

Karpouzas, D. G., Walker, A., Froud-Williams, R. J., and Drennan, D. S. H. (1999), Evidence for the enhanced biodegradation of ethoprophos and carbofuran in soils from Greece and the UK, *Pestic Sci.,* 55: 301–311.

Kasilo, O. J., Hobane, T. and Nhachi, C. F. B. (1991), Organophosphate poisoning in urban Zimbabwe, *J. Appl. Toxicol.,* 11, 269–272.

Kasprzyk-Hordern, B., Ziolek, M. and Nawrocki, J. (2003), Catalytic ozonation and methods of enhancing molecular ozone reactions in water treatment, *Appl. Catal. B: Environ*, 46 (4): 639-669.

Kearney, P. and Wauchope, R. (1998), *Disposal options based on properties of pesticides in soil and water.* In: Kearney P. and Roberts T. (Eds.) *Pesticide remediation in soils and water*, Wiley Series in Agrochemicals and Plant Protection.

Kearney, P. C., Muldoon, M. T. and Somich, C. J. (1987), UV-Ozonation of 11 major pesticides as a waste-disposal pretreatment, *Chemosphere,* 16(10–12):2321–2330.

Keith, L. H. (1997), *Environmental Endocrine Disruptors - A Handbook of Property Data.* JohnWhiley & Sons, Inc., New York.

Keith, L. H. and Telliard, W. A. (1979), Priority pollutants1: A perspective review, *Environ. Sci. Technol.,* 13: 416–423.

Keprasertsup, C., Upatham, E. S., Sukhapanth, N. and Prempree, P. (2001), Degradation of methyl parathion in an aqueous medium by soil bacteria, *Sci Asia.,* 27:261–270.

Khan, B. A., Farid, A., Asi, M. R., Shah, H. and Badshah, A. K. (2009), Determination of residues of trichlorfon and dimethoate on guava using HPLC, *Food Chem.,* 114: 286-288.

Khan, S. A., Hamayun, M., Ahmed, S. (2006), Degradation of 4-aminophenol by newly isolated *Pseudomonas* sp. strain ST-4, *Enzyme Microb Technol.,* 38:10–13.

Khuri, A. I. and Cornell, J. A. (1996), *Response Surfaces*. 2nd ed. New York: Dekker.

Kim, B. S., Fujita, H., Sakai, Y., Sakoda, A. and Suzuki, M. (2002), Catalytic ozonation of an organophosphorus pesticide using microporous silicate and its effect on total toxicity reduction, *Water Sci. Technol*., 46(4-5): 35–41.

Kim, J. R., Ahn, Y. J. (2009), Identification and characterization of chlorpyrifos-methyl and 3, 5, 6-trichloro-2-pyridinol degrading *Burkholderia* sp. strain KR100. *Biodegradation* 20:487–497.

Kim, J. W., Rainina, E., Engler, C., and Wild, J. (1999). Processing efficiency of immobilized non growing bacteria: biocatalytic modeling and experimental analysis, *The Can. J. Chem. Eng*, 77:883-892.

Kim, Y. H., Ahn, J. Y., Moon, S. H. and Lee, J. (2005), Biodegradation and detoxification of organophosphate insecticide, Malathion by *Fusarium oxysporum* f. sp. pisi cutinase, *Chemosphere,* 60(10): 1349–1355.

Kimbara, K., Hashimoto, T., Fukuda, M., Koana, T., Takagi, M., Oishi, M. and Yano, K. (1989), Cloning and sequencing of 2 Tendem genes involved in polychlorinated biphenyl degrading soil bacterium *Pseudomonas* sp. strain KKS102, *J Bacteriol*., 171 (5): 2740–2747.

Kimura, M. (1980), A simple method for estimating evolutionary rates of base substitutions through comparative studies of nucleotide sequences. *J. Mol. Evol.,* 16: 111-120.

Kiran, B., Kaushik, A. and Kaushik, C. P. (2007), Response surface methodological approach for optimizing removal of Cr (VI) from aqueous solution using immobilized cyanobacterium, *Chem Eng.,* J 126: 147-153.

Kitagawa, W., Kimura, N. and Kamagata, Y. (2004), A novel p-nitrophenol degradation gene cluster from a gram-positive bacterium, *Rhodococcus opacus* SAO101, *J Bacteriol.,* 186:4894-4902.

Klimek, M., Lejck, B., Kafarski, P. and Forlani, G. (2001), Metabolism of the phosphonate herbicide glyphosate by a non-nitrateutilising strain of *Penicillium chrysogenum, Pest Mang Sci.,* 57: 815–821.

Kononova, S. V. and Nesmeyanova, M. A. (2002). Phosphonates and their degradation by microorganisms. *Biochemistry (Moscow).* 67:184–195.

Kotronarou, A., Mills, G. and Hoffmann, M. R. (1991), Ultrasonic irradiation of p-nitrophenol in aqueous solution, *J. Phy. Chem*. 95: 3630.

Krishna, K. R., Philip, L. (2008), Biodegradation of lindane, methyl parathion and carbofuran by various enriched bacterial isolates. *J Environ Sci Health B.,* 43:157-171.

Krishnaswamy, U. GCMS and FTIR spectral analysis of aqueous methylparathion biotransformation by the microbial mpd strains of *Pseudomonas aeruginosa* and *Fusarium spp. ArchMicrobiol* (2021). https://doi.org/10.1007/s00203-021-02520-2.

Krzysko-Lupicka, T., Stroff, W., Kubs, K., Skorupa, M., Wieczorek, P., Lejczak, B., and Kafarski, P. (1997), The ability of soil-borne fungi to degrade organophosphonate carbon-to-phosphorus bonds, *Appl. Environ. Microbiol*, 48: 549-552.

Ku, Y., and Lin, H. S. (2002), Decomposition of phorate in aqueous solution by photolytic ozonation, *Wat. Res.*, 36 (16): 4155-4159.

Ku, Y., Chang, J. L., Shen, Y. S. and Lin, S. Y. (1998), Decomposition of diazinon in aqueous solution by ozonation, *Water Res.*, 32:1957–1963.

Kullman S. W. and Matsumura F. (1996), Metabolic Pathways Utilized by *Phanerochaete chrysosporium* for Degradation of the Cyclodiene Pesticide Endosulfan, *Appl. Environ. Microbiol.,* 62(2): 593–600.

Kumar, S., Dudeley, J., Nei, M. and Tamura, K., (2008), A biologist-centric software for evolutionary analysis of DNA and protein sequences, *Briefings in bioinformatics,* 9: 299-306.

Kuo, L. Y. and Perera, N. M. (2000), Paraoxon and Parathion Hydrolysis by Aqueous Molybdenocene Dichloride (Cp2MoCl2): First Reported Pesticide Hydrolysis by an Organometallic Complex. *American Chemical Society, Inorganic Chemistry*. 39(10): 2103-2106.

Kuo, W. S. (2002), Photocatalytic Oxidation of Pesticide Rinsate, J. Environ. Sci. Health Part B -Pestic. Contam. *Agric. Wastes*., 37(1): 65–74.

Lai, K., Stolowich, N. J. and Wild, J. R. (1995), Characterization of P–S bond hydrolysis in organophosphorothioate pesticides by organophosphorus hydrolase, *Arch Biochem Biophys,* 318:59–64.

Laplanche, A., Martin, G. and Tronnard, E. (1984), Ozonation schemes of organophosphorus pesticides: application in drinking water treatment, *Ozone-Sci. Eng.,* 6:207–219.

Laville, N., Balaguer, P., Brion, F., Hinfray, N., Casellas, C., Porcher, J. M., and Ait-Aissa, S. (2006), Modulation of aromatase activity and mRNA by various selected pesticides in the human choriocarcinoma JEG-3cell line, *Toxicology,* 10:228(1):98-108.

Lee, J., Ye, L., Landen, W. O. and Eitenmiller, R. R. (2000), Optimization of an extraction procedure for the quantification of vitamin E in tomato and broccoli using response surface methodology, *J Food Comp. Anal.,* 13, 45-57.

Lee, S. G., Yoon, B. D., Park, Y. H. and Oh, H. M. (1998), Isolation of a novel pentachlorophenol degrading bacterium, *Pseudomonas* sp. Bu 34, *J. Appl. Microbiol,* 85: 1-8.

Leung, K. T., Tresse, O., Errampalli, D., Lee, H. and Trevors, J. T. (1997), Mineralization of p-nitrophenol by pentachloro phenol degrading *Sphingomonas* spp, *FEMS Microbiol. Lett,* 155: 107–114.

Leung, K., Campbell, S., Gan, Y., White, D., Lee, H. and Trevors, J. (1999), The role of the *Sphingomonas* species UG30 pentachlorophenol-4-monooxygenase in p-nitrophenol degradation, *Fems Microbiol. Letters*, 173 (1): 247-253.

Leung, K. T., Tresse, O., Errampalli, D., Lee, H. and Trevors, JT. (1999), Mineralization of p-nitrophenol pentachlorophenol degrading *Sphingomonas* sp, *FEMS Microbiol Lett,* 155:107-114.

Lewis, D. L., Hodson, R. E. and Freeman, L. F. (1984), Effects of microbial community interactions on transformation rates of xenobiotic chemicals, *Appl. Environ. Microbiol,* 48: 561-565.

Li, W., Yun Dai, Beibei Xue, Yingying Li, Xiang Peng, Jingshun Zhang, Yanchun Yan. (2009), Biodegradation and detoxification of endosulfan in aqueous medium and soil by Achromobacter xylosoxidans strain CS5. *J Hazard Mater* 167:209–216.

Liang, W. Q., Wang, Z. Y., Li, H., Wu, P. C., Hu, J. M., Luo, N., Cao, L. X. and Liu, Y. H. (2005), Purification and characterization of a novel pyrethroid hydrolase from *Aspergillus niger* ZD11, *J Agr Food Chem.*, 53: 7415-7420.

Lin, T. S. and Kolattukudy, P. E. (1980), Isolation and characterization of a cuticular polyester (cutin) hydrolyzing enzyme from phytopathogenic fungi, *Physiol Plant Pathol.*, 17: 1-5.

Lipok, J., Dombrovska, L., Wieczorek, P. and Kafarski, P. (2003), *The ability of fungi isolated from stored carrot seeds to degrade organophosphonate herbicides*. Pesticide in Air, Plant, Soil and Water System (Del Re AAM, Capri E, Padovani L &Trevisan M, Eds), Proceeding of the XII Symposium Pesticide Chemistry, Piacenza, Italy.

Liu, C. M., Mclean, P. A., Sookde, C. C. and Cannon, F. C. (1991), Degradation of the herbicide glyphosate by members of the family Rhizobiaceae, *Appl Environ Microbiol.,* 57: 1799–1804.

Liu, G. L. and Lin, Y. (1995), Electrochemical sensor for organophosphate pesticides and nerve agents using Zirconia nanoparticles as selective Sorbents, *Anal. Chem.,* 5894-5901.

Liu, H., Zhang, J., Wang, S., Zhang, X. and Zhou, N. (2005), Plasmid-borne catabolism of methylparathion and p-nitrophenol in *Pseudomonas* sp. strain WBC-3, *Biochem Biophys Res Commun.*, 334: 1107–1114.

Liu, P., Song, X. X., Wen, W. H. and Yuan, W. H. (2006), *Effects of mixed cypermethrin and methylparathion on endocrine hormone levels and immune functions in rats: II.* Interaction, Wei Sheng Yan Jiu*,* 35(5):531-3 [English abstract].

Liu, Y. H., Chung, Y. C. and Xiong, Y. (2001), Purification and characterization of a dimethoate degrading enzyme of *Aspergillus niger* ZHY256, isolated from sewage, *Appl Environ Microbiol.*, 67: 3746–3749.

Liu, Y. H., Liu, Y., Chen, Z. S., Lian, J., Huang, X. and Chung, Y. C. (2004), Purification and characterization of a novel organophosphorus pesticide hydrolase from *Penicillium lilacinum* BP303, *Enzyme. Microbial. Technol.*, 34: 297-303.

Malghani, S., Chatterjee, N., Hu, X. and Zejiao L. (2009), Isolation characterization of a profenofos degrading bacterium, *J Environ Sci.,* 21:1591–1597.

Mallic, K., Bharati, K., Banerji, A., Shakil, N. A. and Sethunathan, N. (1999), Bacterial degradation of chlorpyrifos in pure culture and soil, *Bull. Environ. Contam. Toxicol,* 62: 48-54.

Margesin, R. and Schinner, F. (1997), Effect of temperature on oil degradation by a psychrotrophic yeast in liquid culture and in soil, *FEMS Microbiol., Ecol* 24: 243-249.

Marinho, G., Rodrigues, K., Araujo, R., Pinheiro, Z. B. and Marinho Silva, G. M. (2011), Glucose effect on degradation kinetics of methyl parathion by filamentous fungi species *Aspergillus niger* AN400, *Eng Sanit Ambient*, 16(3): 225-230.

Martin, M., Mengs, G., Plaza, E., Garbi, C., Sanchez, M., Gibello, A., Gutierrez, F. and Ferrer, E. (2000), Propachlor removal by *Pseudomonas* strain GCH 1 in an immobilized cell system, *Appl. Environ. Microbiol*, 66 (3): 1190-1194.

Masten, S. J. and Davies, S. H. R. (1994), The use of ozonation to degrade organic contaminants in wastewaters, *Environ. Sci. Technol*, 28(4): A, 180–A185.

Mathur, S. C. (1999), Future of Indian pesticides industry in next millennium. *Pesticide information,* 24 (4): 9-23.

Matsumura, F. and Boush, G. M. (1966), Malathion degradation by *Trichoderma viride* and a *Pseudomonas* species, *Science,* 153: 127–128.

McAuliffe, K. S., Hallas, L. E. and Kulpa, C. F. (1990), Glyphosate degradation by *Agrobacterium radiobacter* isolated from activated-sludge, *J. Ind. Microbiol*., 6: 219–221.

McDaniel, C. S., Harper, L. L. and Wild, J. R. (1988), Cloning and sequencing of a plasmid-borne gene (opd) encoding a phosphotriesterase, *J Bacteriol.,* 170:2306–2311.

Megharaj, M., Venkateswarlu, K. and Rao, A. S. (1987), Metabolism of monocrotophos and quinalphos by algae isolated from soil, *Bull. Environ, Contam. Toxicol,* 39: 251–256.

Meijers, R. T., Oderwald-Muller, E. J., Nuhn, P. A. N. M. and Kruithof, J. C. (1995), Degradation of pesticides by ozonation and advanced oxidation, *Ozone-Sci. Eng*. 1: 673–686.

Meilgaard, M., Civille, G. V. and Carr, B. T. (2002), *Advanced statistical methods, In Sensory Evaluation Techniques*, second ed. CRC Press, Boca Raton, FL, 275-304.

Meister, R. T. (1992), *Farm Chemicals Handbook '92*. Meister Publishing Company, Willoughby, OH.

Meng, Chun., Chngchun, S. and Guo, Yanghao., (2004), Study on characteristics of bio-cometabolic removal of omethoate by the *Aspergillus* spp, *Wat Res.,* 38:1139-1146.

Mick, D. L. and Dahm, P. A. (1970), Metabolism of parathion by two species of *Rhizobium*, *J. Econ. Entomol,* 63: 1155–1159.

Miller D. F., Alkezweeny A. J., Hales J. M. and Lee R. N, (1978), Ozone formation related to power plant emissions, *Science, N.Y. 202:* 1186.-1188.

Mishra, D., Bhuyan, S., Adhya, T. K. and Sethunathan, N. (1992), Accelerated degradation of methyl parathion, parathion and fenitrothion by suspensions from methyl parathion and p-nitrophenol treated soils, *Soil Biol Biochem.,* 24:1035-1042.

Mitra, D., Vaidyanathan, C. S. (1984), A new 4-nitrophenol 2-hydroxylase from a *Nocardia* sp, *Biochem Int.,* 8:609–615.

Miyamoto, J., Kitagawa, K. and Sato, Y. (1966), Metabolism of organophosphorus insecticides by *Bacillus subtilis* with special emphasis on Sumithion, *Japan J. Expl. Med.,* 36(2): 211-225.

Moghadamnia, A. A. and Abdollahi, M. (2002), *East Mediterr Health J*, 8: 88-94.

Montgomery, D. C. (1991), *Design and Analysis of Experiments,* 3rd ed., Wiley, New York, 1991, p. 270.

Montgomery, D. C. (2004), *Design and analysis of experiments*, 5th edition. New York: Wiley.

Moore, I. K., Braymer, H. D. and Larson, A. D. (1983), Isolation of a *Pseudomonas* sp. which utilizes the phosphonate herbicide glyphosate, *Appl Environ Microbiol.,* 46: 316–320.

Mostafa, I. Y., Bahig, M. R. E., Fakhr, I. M. I. and Adam, Y. (1972a), Metabolism of organophosphorus insecticides. XIV. Malathion breakdown by soil fungi, Z. *Naturforsch,* 27: 1115–1116.

Mostafa, I. Y., Fakhr, I. M. I., Bahig, M. R. E. and El-Zawahry, Y. A. (1972b), Metabolism of organophosphorus insecticides. XIII. Degradation of Malathion by *Rhizobium* spp., *Arch. Microbiol.*, 86: 221–224.

Mulbry, W. W. and Karnas, J. S. (1989a), Parathion hydrolase specified by *Flavobacterium* opd gene: relationship between gene and protein, *J. Bacteriol.,* 171: 6740– 6746.

Mulbry, W. W. and Karns, J. S. (1989), Purification and characterization of three parathion hydrolases from gram-negative bacterial strains, *Appl. Environ. Microbiol,* 55: 289-293.

Mulbry, W. W. and Karns, J. S. (1989b), Purification and characterization of three parathion hydrolases from Gram-negative bacterial strains, *Appl. Environ. Microbiol,* 55: 289–293.

Mulbry, W. W., Karns, J. S., Kearney, P. C., Nelson, J. O., McDaniel, C. S. and Wild, J. R. (1986), Identification of a plasmid-borne parathion hydrolase gene from *Flavobacterium* sp. by southern hybridization with opd from *Pseudomonas diminuta, Appl. Environ. Microbiol*, 51: 926-930.

Mulbry, W. W., Kearney, P. C., Nelson, J. O. and Karns, J. S. (1987), Physical comparison of parathion hydrolase plasmid from *Pseudomonas diminuta* and *Flavobacterium* sp, *Plasmid.,* 18: 173–177.

Munnecke, D. M. (1977), Properties of an immobilized pesticide hydrolyzing enzyme, *Appl. Environ. Microbiol*, 33: 503–507.

Munnecke, D. M., and Hsieh, D. P. H. (1974), Microbial decontamination of parathion and p-nitrophenol in aqueous media, *Appl Environ Microbiol*, 28: 212-217.

Munnecke, D. M. (1976), Enzymatic hydrolysis of organophosphate insecticides, a possible pesticide disposal method, *Appl and Environ Microbiol*., 32: 7-13.

Munnecke. D. M. and Hsieh, D. P. H. (1976), Pathways of microbial metabolism of parathion, *Appl. Environ. Microbiol,* 31: 63–69.

Murphy, C. A., Cameron, J. A., Huang, S. J. and Vinopal, R. T. (1996), *Fusarium* polycaprolactone depolymerase is cutinase, *Appl and Environ Microbiol.,* 62:456-460.

Muthukumar, M., Sargunamani, D., Selvakumar, N. and Venkata Rao, J. (2004), Optimisation of ozone treatment for color and COD removal of acid dye effluent using central composite design experiment, *Dyes Pigments*, 63 (2):127–134.

Myers, R. H. and Montgomery, D. C. (2002), *Response Surface Methodology: Process and product optimization using designed experiments*, 5th Edition, John Wiley, New York.

Namba, T., Nolte, C., and Jackrel, J. (1971), *Am J Med*, 50: 475-492.

Narayana, K., Prashanthi, N., Nayanatara, A., Kumar, H. H., Abhilash, K. and Bairy, K. L. (2006a). Neonatal methyl parathion exposure affects the growth and functions of the male reproductive system in the adult rat, *Folia Morphol (Warsz).,* 65(1):26-33.

Nayak, P, & Solanki, H (2021). Pesticides and Indian agriculture- a review. *International Journal of Research - GRANTHAALAYAH*, *9*(5), 250. doi: 10.29121/granthaalayah. v9.i5.2021.3930.

National Bank for Agriculture Rural Development (NABARD'S) Report, http://www.nabard.org/modelbankprojects/land_development.asp.

Nazarian, A. and Amini, B. (2008), Detection of *Pseudomonas* and *Flavobacterium* species harboring organophosphorus degrading elements from the environment, *Iran. J. Basic. Med. Sci.*, 10: 239-244.

Nelson, L. M. (1982), Biologically induced hydrolysis of parathion in soil: isolation of hydrolyzing bacteria, *Soil Biol Biochem.*, 14: 219-222.

Ningthoujam, D. (2005), Isolation and identification of Brevibacterium linen strain degrading p-nitrophenol, *Afr J Biotechnol.*, 4:256–257.

Nobel, (1980), *Ozone Oxidation capabilities*, Ozomax Ltd written by Amir Salama, 2000. http://www.ozomax.com/pdf/capabilities.pdf.

Obojska, A., Lejczak, B. and Kubrak, M. (1999), Degradation of phosphonates by streptomyces isolates, *Appl Microbiol Biotechnol.*, 51: 872–876.

Obojska, A., Ternana, N. G., Lejczak, B., Kafarski, P. and McMullan, P. (2002), Organophosphate utilization by the thermophile Geobacillus caldoxylosilyticus T20, *Appl Environ Microbiol.,* 68: 2081–2084.

Occupational Health Services, Inc. (1991) (Feb. 25). *MSDS for Methyl Parathion.* OHS Inc., Secaucus, NJ.

Ohashi, N., Tsuchiya, Y., Sasano, H. and Hamada, A. (1993), Screening on the reactivity of organic pesticides with ozone in water and their products, *Jpn. J. Toxicol. Environ. Health,* 39(6): 522–533.

Ohashi, N., Tsuchiya, Y., Sasano, H. and Hamada, A. (1994), Ozonation products of organophosphorous pesticides in water, *Jpn. J. Toxicol. Environ. Health*, 40(2): 185–192.

Ohshiro, K., Kakuta, T., Sakai, T., Hidenori, H., Hoshino, T. and Uchiyama, T. (1996), Biodegradation of organophosphorus insecticides by bacterial isolated from turf green soil, *J Fermen Bioeng.*, 82: 299–305.

Oller, I., Malato, S., Sanchez-Perez, J. A., Maldonado, M. I. and Gasso, R. (2007), Detoxification of wastewater containing five common pesticides by solar AOPs-biological coupled system, *Catal. Today*, 129: 69-78.

Oppenländer, T. (2003), *Photochemical purification of water and air: Advanced oxidation processes (AOPs): principles, reaction mechanisms, reactor concepts,* John Wiley, Inc., ISBN 3527305637, Chichester, U.K.

Ou, L. T. and Sharma, A. (1989), Degradation of methyl parathion by a mixed bacterial culture and a *Bacillus* sp. isolated from different soils, *J Agr Food Chem.*, 37: 1514-8.

Pahm, M. and Alexander, M. (1993), Selecting inocula for the biodegradation of organic compounds at low concentrations, *Microb. Ecol.,* 25: 275-286.

Pakala, S. B., Gorla, P., Pinjari, A. B., Krovidi, R. K., Baru, R., Yanamandra, M., Merrick, M. and Siddavattam, D. (2006), Biodegradation of methyl parathion and p-nitrophenol: evidence for the presence of a p-nitrophenol 2-hydroxylase in a Gram-negative *Serratia* sp. Strain DS001, *Appl. Microbiol. Biotechnol,* 73:1452– 1462.

Palleroni, N. J. (1984), In Krieg NR, Holt JG (eds) *Genus Pseudomonas In Bergey's manual of systematic bacteriology,* Vol I. Baltimore, MD: Williams and Wilkins, 140–182.

Paris, D. F., Lewis, D. L. and Wolfe, N. L. (1975), Rates of degradation of malathion by bacteria isolated from the aquatic system, *Environ. Sci. Technol.*, 9,:135–138.

Penaloza-Vazquez, A., Mena, G. L., Herrera-Estrella, L. and Bailey, A. M. (1995), Cloning and sequencing of the genes involved in glyphosate utilization by *Pseudomonas pseudomallei*, *Appl Environ Microbiol.*, 61: 538–543.

Pesticide News. (1995), *Methyl parathion, Pesticide News*, 36: 20-21. Available at http://www.pan-uk.org/pestnews/actives/methylpa.htm.

Petersen, M. T. N., Martel, P., Petersen, E.I., Drabløs, F. and Petersen, S. B. (1997), *Surface and electrostatics of cutinases. In: Methods in Enzymology* 284, B. Rubin and E. A. Dennis. New York, Academic Press, p.130-154.

Petit, F., Le Goff, P., Cravédi, J. P, Valotaire, Y. and Pakdel, F. (1997), Two complementary bioassays for screening the estrogenic potency of xenobiotics: recombinant yeast for trout estrogen receptor and trout hepatocyte cultures, *J Molec Endocrin.,* 19:321–335.

Phillips, J. P., Xin, J. H., Kirby, K., Milne, C. P., Krell, P. and Wild, J. R. (1990), Transfer and expression of an organophosphate insecticide-degrading gene from *Pseudomonas* in *Drosophila melanogaster*, *Proc. Natl. Acad. Sci.* U.S.A., 87: 8155- 8159.

Phillips, M. R., Li, X. and Zhang, Y. (2002). Suicide rates in China, 1995–99. *Lancet* 359, 835-840.

Pieper, D. H. and Reineke, W., (2000), Engineering bacteria for bioremediation, *Curr. Opin. Biotechnol,* 11(3): 262-270.

Pignatello, J. J. and Sun, Y. (1995), Complete oxidation of metolachlor and methyl parathion in water by the photoassisted Fenton reaction, *Wat. Res.,* 29: 1837-1844.

Pike, R. and Amrhein, N. (1988), Isolation and characterization of a mutant of *Arthrobacter* sp. strain GLP-1 which utilizes the herbicide glyphosate as its sole source of phosphorus and nitrogen, *Appl Environ Microbiol.*, 54: 2868-2870.

Pinheiro, Z. B., Kelly Rodrigues, Carlos Ronald Pessoa-Wanderley, Rinaldo dos Santos Araújo, Glória Marinho. (2010), Remoção biológica de fenol por uso de reator contínuo com inóculo de *Aspergillus niger, Revista Engenharia Sanitária*, 15(1): 47-52.

Pipke, R., Amrhein, N., Jacob, G. S., Kishore, G. M. and Schaefer, J. (1987), Metabolism of glyphosate in an *Arthrobacter* sp, GLP-1, *Eur J Biochem.*, 165: 267-273.

Pitter, P., and Chudoba, (1990), *Biodegradability of organic substances in the aquatic environment*, CRC Press, Boston.

Pivetz, B., Kelsey, J., Steenhuis, T. and Alexander, M. (1996), A procedure to calculate biodegradation during preferential flow through heterogeneous soil columns, *Soil Sci. Soc. of America J.,* 60 (2): 381-388.

Poche, P. and Prados, M. (1995), Removal of pesticides by use of ozone or hydrogen peroxide/ozone, *Ozone Sci. Eng.,* 17: 657–672.

Pointing, S. (2001), Feasibility of bioremediation by white rot fungi, *Appl. Microbiol. and Biotechnol.*, 57: 20-33.

Prenafeta boldu, F. X. (2002), *Growth of on aromatic hydrocarbons: environmental technology perspectives.* The Netherlands: Thesis Wageningen University.

Pulgarin, C., Invernizzi, M., Parra, S., Sarria, V., Polania, R. and Péringer, P. (1999), Strategy for the coupling of photochemical and biological flow reactors useful in mineralization of biorecalcitrant industrial pollutants, *Catal. Today*, 54: 341-352.

Qiu, X. H., Bai, W. Q, Zhong, Q. Z, Li, M., He, F. Q. and Li, B. T. (2006), Isolation and characterization of a bacterial strain of the genus *Ochrobactrum* with methyl parathion mineralizing activity, *J Appl Microbiol.*, 101: 986-994.

Qiu, X., Wu, P., Zhang, H., Li, M. and Yan, Z. (2009), Isolation and characterization of *Arthrobacter* sp. HY2 capable of degrading a high concentration of p-nitrophenol, *Bioresour Technol.*, 100:5243-5248.

Qiu, X., Zhong, Q., Li, M., Bai, W. and Li, B. (2007), Biodegradation of p-nitrophenol by methyl parathion-degrading *Ochrobactrum* sp. B2, *Int Biodeterior Biodegrad.,* 59:297-301.

Quintás, G., Noé, A. M., Armenta, S., Garrigues, S. and Guardia, M. D. (2004), Fourier transform infrared spectrometric determination of Malathion in pesticide formulations, *Analytica Chimica Acta* 502: 213-220.

Racke, K. D. and Coats, J. R. (1988), Comparative degradation of organophosphorus insecticides in soil: specificity of enhanced microbial degradation, *J Agric Food Chem.,* 38: 193-199.

Racke, K. D. and Coats, J. R., (1987), Enhanced degradation of isofenophos by soil microorganisms, *J. Agric. Food Chem.*, 35: 94–99.

Racke, K. D., Skidmore, M. W., Hamilton, D. J., Unsworth, J. B., Miyamoto, J. and Cohen, S. J. (1997), Pesticide fate in tropical soils, *Pure Applied Chem.,* 69: 1349-1371.

Ragnarsdottir, K. V. (2000), Environmental fate and toxicology of organophosphate pesticides, *J Geological Soc,* 157: 859–876.

Ramanand, K., Sharmila, M. and Sethunathan, N. (1988), Mineralization of carbofuran by a soil bacterium, *Appl. Environ. Microbiol*, 54: 2129–2133.

Ramanathan, M. P and Lalithakumari, D. (1999), Complete mineralization of methylparathion by *Pseudomonas* sp. A3, *Appl Biotechnol.*, 80 (1): 1-12.

Rani, N. L., and Lalithakumari, D. (1994), Degradation of methylparathion by *Pseudomonas putida*, *Can. J. of Microbiol*, 40: 1000-1006.

Rao, A. V. and Sethunathan, N. (1974), Degradation of parathion by *Penicillum waksmani Zaleski* isolated from flooded acid sulphate soil, *Arch. Microbial.*, 97: 203-208. (NC).

Raushel, F. M. and Holden, H. M. (2000), Phosphotriesterase: an enzyme in search of its natural substrate, *Advan. Enzymol. Relat. Areas Mol. Biol.,* 74:51–93.

Raveh, L., Y. Segall, H. Leader, N. Rothschild, D. Levanon, Y. Henis, Y. Ashani. (1992), Protection against tabun toxicity in mice by prophylaxis with an enzyme hydrolyzing organophosphate esters, *Biochem. Pharmacol*, 44:397–400.

Ravi, G., Rajendiran, C., Thirumalaikolundusubramanian, P. and Babu, N. (2007), *Poison control, training and research center, Institute of Internal Medicine*, Government General Hospital, Madras Medical College, Chennai, India. Presented at 6th Annual Congress of Asia Pacific Association of Medical Toxicology, Bangkok, Thailand 2-14 December 2007.

Raymond, D. G. M., and M. Alexander. (1971), Microbial metabolism and cometabolism of nitrophenols, *Pestic. Biochem. Physiol.*, 1:123-130.

Rebby, B. R. and Sethunathan, N. (1983), Mineralization of parathion in rice rhizosphere, *Appl. Environ. Microbiol*, 45: 826–829.

Reynolds, G., Graham, N., Perry, R. and Rice, R. G. (1989), Aqueous ozonation of pesticides - A review, *Ozone Sci. Eng*., 11(4): 339–382.

Rice, R. G. (1997), Applications of ozone for industrial wastewater treatment - A review, *Ozone Sci. Eng*., 18(6):477–515.

Richins, R., Kaneva, I., Mulchandani, A. and Chen, W. (1997), Biodegradation of organophosphorus pesticides using surface-expressed organophosphorus hydrolase, *Nature Biotechnol*., 15: 984–987.

Rivas, J., Beltran, F. J., Garcia-Araya, J. F. and Navarrete, V. (2001), Simazine removal from water in a continuous bubble column by O_3 and O_3/H_2O_2, *J Environ Sci and Health B*., 36:809–819.

Rodrigues, K. A., Glória Maria Marinho S. Sampaio, Marcelo Zaiat, Sandra Tédde Santaella (2007), Influência da glicose sobre o consumo de fenol por *Aspergillus niger* AN400 em reatores em batelada, *Revista Engenharia Sanitária.,* 12 (2): 222-228.

Roodveldt, C., and Tawfik, D. S. (2005a), Directed evolution of phosphotriesterase from *Pseudomonas diminuta* for heterologous expression in Escherichia coli results in stabilization of the metal-free state, *Protein Eng Des* Sel, 18:51–58.

Roodveldt, C., and Tawfik, D. S. (2005b), Shared promiscuous activities and evolutionary features in various members of the amidohydrolase superfamily, *Biochemistry,* 44:12728–12736.

Rosenberg, A. and Alexander, M. (1979), Microbial cleavage of various organophosphorus insecticides, *Appl Environ Microbiol.,* 37: 886–891.

Sahoo, B. K., Gowda, V., Ghosh, A., Bose, A. and Pal, T. K. (2009), Statistical Evaluation of HPMC Influence on drug release pattern of an Gastro retentive Floating Drug Delivery System, *Pharmind.,* 71(8):1423-1428.

Sahu J. N., Acharya, J. and Meikap, B. C. (2010), Optimization of production conditions for activated carbons from Tamarind wood by zinc chloride using response surface methodology, *Bioresour Technol*., 10:1974-1982.

Sampaio, G. M. M. S. (2005), *Remoção de metil paration e atrazina em reatores com fungos.* Tese (Doutorado em Hidráulica e Saneamento) Escola de Engenharia de São Carlos, Universidade de São Paulo, São Carlos.

Sanchez Lafuente, C., Furlanetto, S. and Fernandez-Arevalo, M. (2002), Didanosine extended release matrix tablets: Optimization of formulation variables using statistical experimental design, *Int. J. Pharm.* 2002; 237 (5):107-118.

Sanyal, D. and Kulshrestha, G. (2004), Degradation of metolachlor in a crude extract of *Aspergillus flavus, J Environ Sci and Health. B.,* 39: 653–664.

Sardar, D. and Kole, R. K., (2005), Metabolism of chlorpyrifos in relation to its effect on the availability of some plant nutrients in the soil, *Chemosphere,* 46: 506-510.

Sarria, V., Parra, S., Adler, N., Péringer, P., Benitez, N. and Pulgarin, C. (2002), Recent developments in the coupling of photoassisted and aerobic biological processes for the treatment of biorecalcitrant compounds, *Catal. Today,* 76: 301-315.

Scott, C., Pandey, G., Hartley, C. J., Jackson, C. J., Cheesman, M. J., Taylor, M. C., Pandey, R., Khurana, J. L., Teese, M., Coppin, C. W., Weir, K. M., Jain, R. K., Lal, R., Russell,

R. J. and Oakeshott, J. G. (2008), The enzymatic basis for pesticide bioremediation, *Indian J. Microbiol.*, 48: 65-79.

Scott, J. P. and Ollis, D. F. (1995), Integration of chemical and biological processes for water treatment: Review and recommendations, *Environ. Progress.*, 14 (2): 88-103.

Sebastian, J., Chandra, A. K. and Kolattukudy, P. E. (1987), Discovery of a cutinase-producing *Pseudomonas* sp cohabiting with an apparently nitrogen-fixing Corynebacterium sp in the phyllosphere, *J Bacteriol.*, 169: 131-136.

Sehested, K., Holcman, J., Bjergbakke, E. and Hart, E. J. (1984), Formation of ozone in the reaction of O_3 and the decay of the ozonide ion radical at pH 10 - 13, *J Phys Chem.*, 88:269–73.

Serdar, C. and Gibson, D. (1985), Enzymatic hydrolysis of organophosphates: cloning and expression of a parathion hydrolase gene from *Pseudomonas diminuta*, *Biotech*, 3: 567-571.

Serdar, C. M., Gaibson, D. T., Munnecke, D. M. and Lancaster, J. H. (1982), Plasmid involvement in parathion hydrolysis by *Pseudomonas diminuta*, *Appl. Environ. Microbiol*, 44: 246–249.

Serdar, C. M., Murdock, D. C. and Rohde, M. F. (1989), Parathion hydrolase gene from *Pseudomonas diminuta* MG: Subcloning, complete nucleotide sequence and expression of the mature portion of the enzyme in *Escherichia coli*, *Biotechnology*, 7: 1151–1155.

Serge C, F., Amadeo, R., Antonio and Eloy Garcia-Calvo. (2000), Pesticide chemical oxidation: state-of-the-art, *Water Res.,* 34 (2): 366–377.

Seth, M. and Chand, S., (2000), Biosynthesis of tannase and hydrolysis of tannins to gallic acid by *Aspergillus awamori* optimization of process parameters, *Process Biochem.,* 36(1): 39-44.

Sethunathan, N. (1971), *Biodegradation of diazinon in a paddy field as a cause of loss of its efficiency for controlling brown plant hoppers in rice fields*, Pest Articles News Summaries (PANS), 17: 18–19.

Sethunathan, N. and Pathak, M. D. (1972), Increased biological hydrolysis of diazinon after repeated application in rice paddies, *J. Agric. Food Chem.*, 20:586–589.

Sethunathan, N. and Yoshida, T. (1973), A *Flavobacterium* that degrades diazinon and parathion, *Can. J. Microbiol.*, 19: 873–875.

Sharmila, M., Ramanand, K. and Sethunathan, N. (1989), Effect of yeast extract on the degradation of organophosphorus insecticides by soil enrichment and bacterial cultures, *Can J Microbiol.*, 35: 1105–1110.

Shen, G. X., Ke, F. Y. and Zhang, J. Q. (2002), Investigation and evaluation of the contents of pollutants in greenhouse vegetables in Shanghai, *Shanghai Environ Sci.*, 21:475-477.

Shen, Y. J., Lu, P., Mei, H., Yu, H. J., Hong, Q. and Li, S. P. (2010), Isolation of a methylparathion degrading strain *Stenotrophomonas* sp. SMSP-1 and cloning of the ophc2 gene, *Biodegradation,* 21:785–792.

Sherine O. Obare, Chandrima De, Wen Guo, Tajay L. Haywood, Tova A. Samuels, Clara P. Adams, Noah O. Masika, Desmond H. Murray, Ginger A. Anderson, Keith Campbel and Kenneth Fletcher. (2010), Fluorescent Chemosensors for Toxic

Organophosphorus Pesticides: A Review, *Sensors,* 10, 7018-7043; doi: 10.3390/s100707018.

Shevchenko, M. A., Taran, P. N. and Marchenko, P. V. (1982), Technology of water treatment and demineralization. Modern methods for purifying water from pesticides, *Soviet J. Wat. Chem. Technol.*, 4(4):53–71.

Shimazu, M., Mulchandani, A. and Chen, W. (2001), Simultaneous degradation of organophosphorus pesticides and p-nitrophenol by a genetically engineered *Moraxella* sp. with surface-expressed organophosphorus hydrolase, *Biotechnology, and Bioeng.*, 76(4): 318-323.

Shroff, R. (2000), *Chairman Address in Pesticide Information,* Annual Issue, New Delhi: Pesticide Association of India Publication.

Siddaramappa, R., Rajaram, K. P. and Sethunathan, N. N. (1973), Degradation of parathion by bacteria isolated from flooded soil, *Appl Microbiol.*, 26: 846–849.

Silva, I. E. C. et al. (2007), Fungos filamentoso degradadores de compostos fenólicos de água residuária de postos de combustíveis, *Biol. Health J.,* 1(1): 123-130.

Simpson, J. R. and Evans, W. C. (1953), The metabolism of nitrophenols by certain bacteria, *Biochem. J.*, 55: 24.

Sing, B. K. and Walker, A. (2006), Microbial degradation of organophosphorus compounds, *FEMS Microbiol Rev.,* 30:428-471.

Singh, A. K. and Seth, P. K. (1989), Degradation of Malathion by microorganisms isolated from industrial effluents, *B. Environ. Contam. Tox,* 43: 28–35.

Singh, B. K., Walker, A., Morgan, A. W., Denis, J. and Wright, D. J. (2004), Biodegradation of chlorpyrifos by *Enterobacter* strain b-14 and its use in bioremediation of contaminated soils, *Appl. Environ. Microbiol*, 70: 4855-4863.

Singh, B. K., Walker, A., Morgan, J. A. W. and Wright, D. J. (2003a), Effects of soil pH on the biodegradation of chlorpyrifos and isolation of a chlorpyrifos degrading bacterium, *Appl. Environ. Microbiol*, 69:5198-5206.

Singh, B. K., Walker, A., Morgan, J. A. W. and Wright, D. J. (2003b), Role of soil pH in the development of enhanced degradation of fenamiphos, *Appl. Environ. Microbiol,* 69: 7035-7043.

Singh, B. K., Kuhad, R. C., Singh, A., Lal, R. and Tripathi, K. K. (1999), Biochemical and molecular basis of pesticide degradation by microorganisms, *Crit. Rev. Biotechnol,* 19: 197–225.

Singh, D. K. and Subhas. (2002), *Biodegradation of monocrotophos by two bacterial isolates.* In: National Conference on Soil Contamination and Biodiversity, February 8-10 Lucknow, 37.

Singh, H. *Mycorremediation.* New Jersey: John Wiley & Sons, 2006. p.592.

Singh, R., Sharma, R., Tewari, N., Geetanjali and Rawat D. S. (2006), Nitrilase and its application as a 'green' catalyst, *Chem. Biodiv.*, 3: 1279-1287.

Singh, S. and Singh, G. K. (2003). Utilization of monocrotophos as phosphorus source by *Pseudomonas aeruginosa* F10B and *Clavibacter michiganense* sub sp. *inisidiosum* SBL 11, *Can J Microbiol.*, 49: 101–109.

Singh, S. K., Dodge, J., Durrani, M. J. and Khan, M. A. (1995), Optimization and characterization of controlled release pellets coated with experimental latex Anionic drug, *Int. J. Pharm.*, 125:243-255.

Slaoui, M., Ouhssine, M., Berny, E. and Elyachioui, M. (2007), Biodegradation of the carbofuran by a fungus isolated from treated soil, *Afr. J. Biotechnol.*, 6(4): 419-423.

Smelt, J. H., Crum, S. J. H., Teunissen, W. and Leistra, M. (1987), Accelerated transformation of aldicarb, oxamyl, and ethoprophos after repeated soil treatments, *Crop Prot,* 6:295–303.

Smelt, J. H., Van de Peppel-Groen, A. E., Van der Plas, L. J. T. and Dijksterhuis, A. (1996), Development and duration of accelerated degradation of nematicides in different soils, *Soil Biol. Biochem*, 28: 1757–1765.

Sogorb, M. A. and Vilanova, E. (2002), Enzymes involved in the detoxification of organophosphorus, carbamate and pyrethroid insecticides through hydrolysis, *Toxicol Lett.,* 128:215–228.

Sogorb, M. A., Vilanova, E. and Carrera, V. (2004), Future application of phosphotriesterases in the prophylaxis and treatment of organophosphorus insecticide and nerve agent poisoning, *Toxicol Lett*., 151: 219- 233.

Somara, S. and Siddavattam, D. (1995), Plasmid-mediated organophosphate pesticide degradation by *Flavobacterium balustinum*, *Biochem Mol Biol Int.,* 36:627–631.

Somara, S., Manavathi, B., Tebbe, C. C. and Siddavatam, D. (2002), Localization of identical organophosphorus pesticide degrading (opd) genes on genetically dissimilar indigenous plasmids of soil bacteria: PCR amplification, cloning and sequencing of opd gene from *Flavobacterium balustinum*, *Indian J Experi Biol.,* 40: 774-779.

Somasundaram, D. J. and Rajadurai, S. (1995), *Acta Psychiatr Scand*., 91: 1-4.

Spain, J. C. and Gibson, D. T. (1991), Pathway for biodegradation of p-nitrophenol in a *Moraxella* sp, *Appl. Environ. Microbiol*, 57:812-819.

Spain, J. C. and Nishino, S. F. (1987), Degradation of 1, 4-dichlorobenzene by a *Pseudomonas* sp, *Appl. Environ. Microbiol*, 53:1010–1019.

Spain, J., Van Veld, P., Monti, C., Pritchard, P. and Cripe, C. (1984), Comparison of p nitrophenol biodegradation in field and laboratory test systems, *Appl. Environ. Microbiol*, 48:944-950.

Stackhouse, S. C. (1980), Determination, isolation and characterization of SD9129-metabolizing microorganisms isolated from freshly collected and pretreated sandy loam soil, *Shell Report RIR*., 22:013-80.

Staehelin, J. and Hoign`e, J. (1982), Decomposition of ozone in water: rate of initiation by hydroxide ions and hydrogen peroxide, *Environ. Sci. Technol*., 16:676–81.

Steiert, J. G., Pogell, B. M., Speedie, M. K. and Laredo, J. A. (1989), A gene coding for membrane-bound hydrolase is expressed as a soluble enzyme in *Streptomyces lividans*, *Biotechnology*., 7: 65–68.

Stiriling, A. M., Stiriling, G. R. and Macrae, I. C. (1992), Microbial degradation of fenamiphos after repeated application to a tomato-growing soil, *Nematologica*, 38: 245–254.

Subhas, D. and Dileep, K. S. (2003), Utilization of monocrotophos as phosphorus source by *Pseudomonas aeruginosa* F10B and *Clavibacter michiganense* subsp. *insidiosum* SBL11, *Can J Microbiol.,* 49(2): 101–109.

Subhas, D. and Dileep, K. S. (2006), Purification and characterization of phosphotriesterase from *Pseudomonas aeruginosa* F10B and *Clavibacter michiganense* subsp. *insidiosum* SBL11, *Can J Microbiol*., 52(2): 157–168.

Subhas, S. and Singh, D. K. (2003), Utilization of monocrotophos as phosphorus source by *Pseudomonas aeruginosa* F10B and *Clavibacter michiganense* subsp. insidiosum SBL 11, *Can. J. Microbiol.*, 49: 101–109.

Subramanian, G., Sekar, S. and Sampoornam, S. (1994), Biodegradation and utilization of organophosphorus pesticides by cyanobacteria, *Int Biodeterior Biodegrad.*, 33: 129–143.

Suett, D. L. and Jukes, A. A. (1997), The Accelerated biodegradation of phorate in carrot soils in the United Kingdom, *Crop Prot.*, 16: 457–461.

Suett, D. L., Fournier, J. C., Papadopoulou-Mourkidou, E., Pussemier, L. and Smelt, J., (1996a), Accelerated degradation: the European dimension, *Soil Biol. Biochem.*, 28: 1741–1748.

Suett, D. L., Jukes, A. A. and Parekh, N. R. (1996b), Non-specific influence of pH on microbial adaptation and insecticide efficacy in previously treated field soils, *Soil Biol. Biochem.*, 28: 1783–1790.

Suett, D. L., Jukes, A. A. and Phelps, K. (1993), Stability of accelerated degradation of soil-applied insecticides. laboratory behavior of aldicarb and carbofuran in relation to their efficacy against cabbage root fly (*Delia radicum*) in previously treated field soils, *Crop Prot.*, 12: 431– 442.

Supak, J. M. (1990), *Genetically engineered pesticide biodegradation by Gliocladium virens*, M.S. Thesis., Texas A & M University, College Station.

Sureshkumar, C., Anuradha, C. M. and Dayananda, S. (1998), *Detoxification of organophosphorus pesticides by bacterial phosphotriesterase.* In 39th AMI Conference, December 5-7, Jaipur, p.236.

Survey, U. S. G. (2010), *Organophosphorus pesticides occurrence and distribution in surface and groundwater of the United States*, Available online: http://ga.water.usgs.gov/publications/ofr00-187.pdf. (Accessed on 26 February 2010).

Tchelet, R., Levanon, D., Mingelrin, D. and Henis, Y. (1993), Parathion degradation by a *Pseudomonas* sp. and a *Xanthomonas* sp. and by their crude enzyme extracts as affected by some cations, *Soil Biol Biochem.*, 25: 1665–1671.

Tebbe, C. C. and Reber, H. H. (1988), Transformation of the herbicide [14C]-glufosinate in soils, *J. Agric. Food Chem.*, 37: 267–271.

Thompson, J. D., Gibson, T. J., Plewniak, F. and Higgians, D. G., (1997), The clustalx windows interface: flexible strategies for multiple sequence alignment added by quality analysis tools. *Nucleic acids research*, 24:4876-4882.

Timmis, K. N. and Pieper, D. H. (1999), Bacteria designed for bioremediation, Trends Biotechnol, 17:201–4.

Topp, E., Crawford, R. L. and Hanson, R. S. (1988), Influence of readily metabolizable carbon on pentachlorophenol metabolism by a pentachlorophenol-degrading *Flavobacterium* sp, *Appl. and Environ. Microbiol*, 54(10):2452-2459.

Tse, H., Comba, M. and Alaee, M. (2004), Methods for the determination of organophosphate insecticides in water, sediments and biota, *Chemosphere*, 54: 41–47.

Tsiropoulos, N. G., Lykas, D. T. and Karpouzas, D. G. (2005), Liquid chromatographic determination of fosthiazate residues in environmental samples and application of the method to a fosthiazate field dissipation study, *J AOAC Inter.*, 88 (6): 1827-1833

U.S. EPA. (2010), *Pesticides and food: Why children may be especially sensitive to pesticides*, Available online: http://www.epa.gov/pesticides/food/pest.htm.

U.S. EPA. 2000. *Pesticides Industry Sales and Usage*: 1996 and 1997 Market Estimates. Office of Pesticide Programs, Available online:

Uesugi, Y. and Tomizawa, C. (1971), Metabolism of O-ethyl S,S Dimethyl phosphorodithiolate (Hinosan) by mycelial cells of Pyricularia oryzae, *Agric. Biol. Chem.,* 35: 941–949.

UNEP/FAO/RC/CRC.1/19/Add.4, (2005), Report and proposed decision of Italy made to the Europe Commission under 91/414/EEC, Vol.1.March 2001.1928/BBA/ECCO/01. http://www.pic.int/incs/crc1/s19add4.

US Census Bureau., Report of the National Intelligence Council's (2020), www.foia.cia.gov/2020/2020.pdf.

US EPA Pesticide Industry Sales and Usage: 2000 and 2001 Market Estimates, updated Jan 2009. Available, http://www.epa.gov/oppbead1/pestsales/01pestsales/market estimates 2001.

US EPA. 2003. Interim Reregistration Eligibility Decision for Methyl Parathion. Case No.0153. United States Environmental Protection Agency. http://www.epa.gov/oppsrrd/REDs/methylparathion.pdf.Signed05/2003. Accessed 20/11/04.

USEPA (1988a) "*Bioremediation of Contaminated Surface Soils,*" EPA/600/2-89/073.

USEPA (1988b) "*Determination of Aerobic Degradation Potential for Hazardous Organic Constituents in Soil- Interim Protocol*, Risk Reduction Engineering Laboratory, Cincinnati, OH.

USEPA (1994a) "*Augmented In-situ Subsurface Bioremediation Process-Demonstration Bulletin,*" EPA/540/MR-93/527, RREL.

USEPA (1994b) "*Ex-situ Anaerobic Bioremediation System-Demonstration Bulletin,*" EPA/540/MR- 94/508, RREL.

Usharani, K., Muthukumar, M. and Kadirvelu, K. (2012), Effect of pH on the Degradation of Aaqueous Organophosphate (Methylparathion) in Wastewater by Ozonation, *In J. Environ Res.,* Spring., 6(2): 557- 564.

Usharani, K., Muthukumar, M. (2013). Optimization of aqueous methylparathion biodegradation by Fusarium sp in batch scale process using response surface methodology. *In. J. Environ. Sci. Technol.*, Springer, 10 (3), 591-606.

Usharani, K., Muthukumar, M. (2016). Aqueous methylparathion removal by Ozonation and Optimization of variables using Central Composite Design of Experiments. *Current Environ. Engg*. 3 (3), 249-266.

Usharani, K., Muthukumar, M. and Kadirvelu, K. (2010), *Effect of pH on the degradation of aqueous organophosphate (methylparathion) in wastewater by ozonation process.* Proceedings of International Conference on Advanced Oxidation Process- ICAOP-2010, Organised by Centre for Environment Education & Technology (CEET) and association with Advanced Centre for Environmental Studies and Sustainable Development (ACESSD) Mahatma Gandhi University, Kottayam, Kerala, India. September, 18-21.

Usharani, K., Muthukumar, M. and Kadirvelu, K. (2010), *Laboratory application of Ozonation process in the treatment of organophosphate contaminated wastewater".* Proceedings of 3rd International Conference on Hydrology and Watershed

Management, Organized by Centre for Water Resources, Institute of Science and Technology, Jawaharlal Nehru Technological University, Hyderabad (AP), India. BSP Publications, www.bspublications.net, ISBN: 978-81-7800-224-8., VOL. II. Pp-1221-1231. Copyright © 2010.

Usharani, K., Muthukumar, M. and Lakshmanaperumalsamy, P. (2006). *Proceedings of Degradative Capacity of Fungi for effective reclamation of Parathion contaminated soils.* 47th AMI -Annual Conference Association of Microbiologists of India, Barkatullah University, Bhopal, India.

Usharani, K., Muthukumar, M. and Lakshmanaperumalsamy, P. (2007). *Biodegradation of para-nitrophenol by methylparathion degrading fungal strain MPD.* Proceedings of 48th AMI -Annual Conference, Association of Microbiologists of India, Indian Institute of Technology, Chennai, TN, India.

Usharani, K., Muthukumar, M. and Lakshmanaperumalsamy, P. (2013), Optimization of phosphate removal from synthetic wastewater by bacterial consortium using Box-Behnken Design, *Environ Eng Manag J., GATU, Romania, 12 (12),* 2371-2383.

Usharani, K., P Lakshmanaperumalsamy. (2016). Box-Behnken experimental design mediated optimization of aqueous methylparathion biodegradation by Pseudomonas aeruginosa mpd strain. *Journal of Microbiology Biotechnology Food Sciences* 5 (6), 534-547.

Usharani Krishnaswamy, (2021) GCMS and FTIR spectral analysis of aqueous methylparathion biotransformation by the microbial mpd strains of *Pseudomonas aeruginosa* and *Fusarium spp. ArchMicrobiol* (2021). https://doi.org/10.1007/s00203-021-02520-2.

Usharani, K., and Arunkumar, V. (2023). Bioremoval and resource recovery of nutrients by phytoremediation using aquatic free-floating plants Lemna minor. *Ukrainian Journal of Ecology*, *13*(5).

Vandana, P. S. and Phale, P. S. (2005), Metabolism of Carbaryl via 1,2 Dihydroxynaphthalene by soil isolates *Pseudomonas* sp strains C_4, C5 and C_6, *Appl. Environ. Microbiol,* 71: 5951-5956.

Veltri, J. and Litovitz, T. (1984), Annual report of the American Associations of poison control centers national data collection system, Am J Emerg Med, 2: 420-443.

Vermeire, T., MacPhail, R. and Waters, M. (2003), Integrated human and ecological risk assessment: A case study of organophosphorous pesticides in the environment. Hum. Ecol. Risk. Assessment, 9: 343-357.

Vlyssides, A., Arapoglou, D., Mai S. and Barampouti, E. M. (2005), Electrochemical detoxification of four phosphorothioate obsolete pesticides stocks, *Chemosphere*, 58: 439–447.

Vlyssides, A., Barampouti, E. M., Mai, S., Arapoglou, D. and Kotronarou, A. (2004), Degradation of methylparathion in aqueous solution by electrochemical oxidation, *Environ. Sci. Technol*., 38 (22): 6125-6131.

Von Gunten, U. (2007), The basics of oxidants in water treatment. Part B: Ozone reactions, *Water. Sci. Technol*, 55 (12): 25–29.

Walker, A. W. and Keasling, J. D. (2002), Metabolic engineering of *Pseudomonas putida* for the utilization of parathion as a carbon and energy source, *Biotechnol Bioeng*., 78:715–721.

Walker, B. J. (1972), *Organophosphorus Chemistry*. Penguin: London, UK.

Walker, W. W. and Stojanovic, B. J. (1974), Malathion degradation by an *Arthrobacter* species, *J. Environ. Qual.,* 3: 4–10.

Wang, H., Marjomaki, V., Ovod, V. and Kulomaa, M. (2002), Subcellular localization of pentachlorophenol 4-monooxygenase in *Sphingobium chlorophenolicum* ATCC 39723, *Biochem. and Biophysical Res. Communications.*, 299(5): 703-709.

Wang, L., Wen, Y., Guo, X., Wang, G., Li, S. and Jiang, J. (2010), Degradation of methamidophos by *Hyphomicrobium* species MAP-1 and the biochemical degradation pathway, *Biodegradation*, 21:513–523.

Wang, S., Zhang, C. and Yan, Y. (2011), Biodegradation of methyl parathion and p-nitrophenol by a newly isolated *Agrobacterium* sp. strain Yw12, *Biodegradation,* 23(1):107-116 (doi10.1007/s10532-011-9490-0).

Watkins, L. M., Mahoney, H. J., McCulloch, J. K., Raushel, F. M. (1997), Augmented hydrolysis of diisopropyl fluorophosphate in engineered mutants of phosphotriesterase, *J. Biol. Chem.*, 272:25596 -25601.

Webster, L. R., Mckenzie, G. H. and Moriarty, H. T. (2002), Organophosphate-based pesticides and genetic damage implicated in bladder cancer, *Cancer Genet Cytogen.,* 133: 112–117.

WHO in collaboration with UNEP, 1990. *Public Health Impact of Pesticides Used in Agriculture.* Updated June 2007, WHO, Geneva.

WHO, 2002. *The World Health Report 2002. Reducing risks, promoting healthy life.* WHO, Geneva.

Wilmes, R. (1987), *Parathion-methyl: Hydrolysis studies, Leverkusen*, Germany, Bayer AG, Institute of Metabolism Research, (Unpublished report No. PF 2883, submitted to WHO by Bayer AG), 34.

Witteveen, Cor F. B. (1993), *Gluconato formation and polyol metabolism in Aspergillus niger.* Thesis Wageningen University, Wagenningen, The Netherlands.

World Health Organization (WHO), (2009), *Health implications from monocrotophos use: a review of the evidence in India*, pp: 1-6.

Wu, C. and Linden, K. G. (2008), Degradation and by-product formation of parathion in aqueous solutions by UV and UV/H_2O_2 treatment, *Water Res.*, 42: 4780-4790. Wu, J., Luan, T., Lan, C., Lo, T. W. H. and Chan, G. Y. S. (2007), Removal of residual pesticides on vegetable using ozonated water, *Food Control*, 18: 466–472.

Wu, J., Rudya, K. and Sparka, J. (2000), Oxidation of aqueous phenol by ozone and peroxidase, *Adv. Environ. Res.,* 4 (4):339-346.

Xu, B., Wild, J. R. and Kernerley, C. M. (1996), Enhanced expression of the bacterial gene for pesticide degradation in a common soil fungus, *J Ferment Bioeng.*, 81: 473–481.

Xu, G., Li, Y., Zheng, W., Peng, X., Li, W. and Yan, Y. (2007), Mineralization of chlorpyrifos by co-culture of *Serratia* and *Trichosporon* spp. *Biotechnol Lett.*, 29:1469–147.

Xu, L., Resing, K., Lawson, S., Babbitt, P. and Copley, S. (1999), Evidence that pcpA encodes 2,6-dichlorohydroquinone dioxygenase, the ring cleavage enzyme required for pentachlorophenol degradation in *Sphingomonas chlorophenolica* strain ATCC 39723., *Biochemistry.*, 38 (24): 7659-7669.

Xu, L., Resing, K., Lawson, S., Babbitt, P. and Copley, S. (1999), Evidence that pcpA encodes 2,6-dichlorohydroquinone dioxygenase, the ring cleavage enzyme required for pentachlorophenol degradation in *Sphingomonas chlorophenol* strain ATCC 39723, *Biochemistry.,* 38 (24): 7659-7669.

Yali, C., Xianen, Z., Hong, L., Yinshan, W., and Xiangming, X. (2002), Study on *Pseudomonas* sp. WBC-3 capable of complete degradation of methyl parathion, *Weishengwu Xuebao.*, 42: 490–497.

Yang O., Li C., Li H., Li Y. and Yu N., (2009), Degradation of synthetic reactive azo dyes and treatment of textile wastewater by a fungi consortium reactor, *Biochem Engg. J.*, 43: 225-230.

Yang, C., Dong, M., Yuan, Y., Huang, Y., Guo, X., and Qiao, C. (2007), Reductive transformation of parathion and methylparathion by *Bacillus* sp, *Biotechnol Lett.,* 29:487-493.

Yang, C., Liu, N., Guo, X. and Qiao, C. (2006), Cloning of mpd gene from a chlorpyrifos degrading bacterium and use of this strain in bioremediation of contaminated soil, *FEMS Microbiol Lett.,* 265:118-125.

Yang, F. C, Huang, H. C. and Yang, M. J (2003), The influence of environmental conditions on the mycelial growth of *Antrodia cinnamomea* in submerged cultures, *Enzyme Microb Technol.*, 33:395–402.

Yang, L., Zhao, Y.-H., Zhang, B.-X., Yang, C.-H. and Zhang, X. (2005), Isolation and characterization of a chlorpyrifos and 3,5,6-trichloro-2-pyridinol degrading bacterium, *FEMS Microbiol. Lett.,* 251: 67-73.

Yang, W., Ya-Feng Zhou, He-Ping Dai, Li-Jun Bi, Zhi-Ping Zhang, Xiao-Hua Zhang, Yan Leng, Xian-En Zhang. (2008), Application of methyl parathion hydrolase (MPH) as a labeling enzyme. *Anal. Bioanal. Chem.,* 390, 2133-2140.

Yang, Y. C. (1995), Chemical reactions for neutralizing chemical warfare agents, *Chem. Ind.*, 8:334- 337.

Yasuno, M., Hirakoso, S., Sasa, M. and Uchida, M. (1965), Inactivation of some organophosphorus insecticides by bacteria in polluted water, *Japan J. of Expl. Med.,* 35(6): 563.

Ye, M. M., Chen, Z. L., Liu, X. W., Ben, Y. and Shen. J. M. (2009), Ozone enhanced activity of aqueous titanium dioxide suspensions for photodegradation of 4-chloronitrobenzen, *J Hazard Mater.,* 167:1021-1027.

Yu, C. P. and Yu, Y. H. (2001), Mechanisms of the reaction of ozone with p-nitrophenol, *Ozone- Sci. Eng.,* 23(4):303-312.

Yugui, T., Yaoming, W., Shilei, Y. and Lianbin, Y. (2008), Optimization of omethoate degradation conditions and a kinetics model, *Int Biodeterior & Biodegrad.*, 62: 239-243.

Yusuf, H. R., Akhter, H. H., Rahman, M. H., Chowdhury, M. K. and Rochat, R. W. (2000), Injury-related deaths amongst women aged 10–50 years in Bangladesh, 1996–97, *Lancet,* 355: 1220–1224.

Zaidi, B. and Mehta, N. (1995), Effects of organic compounds on the degradation of pnitrophenol in the lake and industrial waste-water by inoculated bacteria, *Biodegradation*, 6(4): 275-281.

Zboinska, E., Lejczak, B. and Kafarski, P. (1992a), Organophosphate utilization by the wild-type strain of *Pseudomonas fluorescens*, *Appl Environ Microbiol.,* 58: 2993-2999.

Zhang C. and Bennett G. N. (2005), Biodegradation of xenobiotics by anaerobic bacteria, *Appl Microbiol Biotechnol* 67, 600–618.

Zhang, C., Le Jia, Shenghui Wang, Jie Qu, Kang Li, Lili Xu, Yanhua Shi, Yanchun Yan (2010), Biodegradation of beta-cypermethrin by two Serratia spp. with different cell surface hydrophobicity. *Bioresour Technol* 101:3423–3429.

Zhang, Z., Hong, Q., Xu, J., Zhang, X., and Li, S. (2006), Isolation of fenitrothion-degrading strain Burkholderia sp. FDS-1 and cloning of mpd gene. *Biodegradation,* 17:275–283.

Zhao, G., Li, H. and Niu, S. (2007), Technologic parameter optimization of the gas quenching process using response surface method, *Computat Mater Sci.*, 38 (4): 561-570.

Zhao, W. R., Shi, H. X. and Wang, D. H. (2004), Ozonation of cationic red X-GRL in aqueous solution: degradation and mechanism, *Chemosphere,* 57 (9):1189–1199.

Zheng, W. and Liu, W. (2002), Kinetics and mechanism of the hydrolysis of imidacloprid. *Chinese Chem Letts.,* 13 (12):1170-1173.

Zhongli, C., Ruifu, Z., Jian, H. and Shunpeng, L., (2002), Isolation and characterization of a p-nitrophenol degradation *Pseudomonas* sp. strain p3 and construction of a genetically engineered bacterium, *Weishengwu Xuebao*., 42: 19-26.

Zhongli, C., Shunpeng, L. and Guoping, F. (2001), Isolation of methyl parathion-degrading strain M6 and cloning of the methyl parathion hydrolase gene, *Appl Environ Microbiol.*, 67(10):4922-4925.

Zuckerman, B. M., Deubert, K., Mackiewicz, M. and Gunner, H. (1970), Studies on the biodegradation of parathion, *Plant Soil.,* 33: 273-281.

Zwiener, C., Weil, L. and Niessner, R. (1995), Atrazine and parathion-methyl removal by UV and UV/O_3 in drinking water treatment, *Int. J. Environ. Anal. Chem.*, 58: 247-264.

About the Authors

Dr. Usharani Rathinam Krishnaswamy
Department of Environmental Sciences, Bharathiar University, TN, India
Post-Doctoral Researcher (Young Talent JTEE CAPES Research Fellow,)
CAPES-Higher Education, Ministry of Higher Education, Govt. of Brazil,
Department of Civil and Environmental Engineering,
UNESP, Sao Paulo State University, Brazil,
Email: usharaniks2003@yahoo.com
usharani.rathinam-krishnaswamy@unesp.br

Dr. Usharani RK received her MSc., M.Phil., PhD., (Environmental Sciences specialized in Environmental Microbiology), PG Dip. (Microbial Biotechnology, Bioinformatics), Bharathiar University, IN., Additional Qualifications in Biomedical Nanotechnology, Membrane Technology, Environmental Engineering & Pollution Abatement, Food Microbiology & Safety courses from National Programme on Technology Enhanced Learning-Indian Institute of Technology and Swayam, MHRD, Govt. of India. She also received a course completed on NBS-Nature based solutions for creating circular cities from the United Nations Development Programme: Stockholm, SE. She was the recipient of UGC-NET and qualified for the Indian University Grant Commission-National Eligibility Test, having Research experience (10yrs) in India and Overseas in various organizations, Researcher in DRDO-BU Center for Life Sciences, Bharathiar University, Research Associate in R&D Biotech Lab Pvt. Ltd., Post-Doctoral Researcher in Environmental Engineering and Technology (abroad), and Teaching experience (6yrs) as assistant professor at Bharathiar University affiliated colleges and Central University of Kerala. She has published 30 papers for international peer-reviewed scientific journals, more than 30 international and national conferences, four book chapters for scientific renowned publishers, Springer, Taylor & Francis, and CRC-Press in her credit, and published two books with scientific academic publishers. She received SRj, JRF, and fellowships from

BU-DRDO Center for Life Sciences, Bharathiar University, and the Ministry of Defence, Govt. of India, as well as from the TN State Council Science and Technology student's project fellowship scheme. She was awarded best paper at the International Conference on Advanced Oxidation-ICAOP-2010. She was the recipient of the prestigious Brazil Young Talent JTEE Researcher and Post-Doctoral Research Fellowship from CAPES-Higher Education, Ministry of Higher Education, Govt. of Brazil. She acknowledged and received the Project Pitch awarded for the most holistic and scientifically grounded approach! NBS Edu WORLD, Urban by Nature, European Union., Hackathon, 2024., Sustainable Ambassador 2024, SPSC Ambassador, SPSC, Sustainability Service, UK, and Bharat Shiksha Gaurav Puraskar Award, 2024, for remarkable contribution in the field of education received from KTK outstanding achievers and education foundation, New Delhi, India. She acts as a referee, reviewer for scientifically reputed several journals of Springer, Elsevier, ACS, Taylor and Francis, Bentham Science scientific publishers and editorial board member of scientific frames in the fields of microbiology, biotechnology, environmental engineering, nanotechnology, ASM, EFB, WAST, microbiologist associations, etc. for scientific committees, academic, research, in various organizations, forums and scientific collaboration with international institutions.

Orcid: *https://orcid.org/0000-0003-4077-2497*
Google Scholar:
https://scholar.google.com/citations?user=CDkddu8AAAAJ&hl=en
SCOPUS ID:
https://www.scopus.com/authid/detail.uri?authorId=55060416300
Researchghate: *https://www.researchgate.net/profile/Usharani-K*
CNPq, Brazil: http://lattes.cnpq.br/2184965895317068
Research ID: *https://researchid.co/usharani*

Recent Publications

2024. Sustainable Removal of Pollutants and Recovery of Nutrients from wastewater by Ecotechnological Approaches using Cohesive Phyto-Microbial-Electrochemical processes. *5th International Conference-Strategies toward Green Deal Implementation Water, Raw Materials & Energy in Green Transition*, 27-29 November 2024, Poland.

2024. Biochar and Nanobiochar: A Biobased Engineered Nanobiocatalyst from Sugarcane Bagasse in Petrochemical Waste Management. *12th Brazilian Congress of Research and Development in Oil and Gas*, Brazil. October 30 to November 01, 2024, Brazil.

2023. Bioremoval and Resource recovery of Nutrients by Phytoremediation using Aquatic Free-Floating Plants Lemna minor, *Ukrainian Journal of Ecology.*, 13 (5), 13-27; ISSN: 2520-2138 Melitopol: Alex Matsyura Press.

2022. Ecotechnological Remediation: Application of aquatic free-floating plants (Lemna minor) in the Phytoremediation of Iron, Nitrate, Phosphate from wastewater. *International Conference on Advances in Energy, Environment for Sustainable Development,* 7-8th January, 2022. Organized by Dept. of Chemistry Siksha 'O' Anusandhan (Deemed to be University), Odisha, India and In collaboration with Dept. of Civil Engineering, NIT, Meghalaya, India.

2021. GCMS and FTIR spectral analysis of aqueous methylparathion biotransformation by the microbial mpd strains of Pseudomonas aeruginosa and Fusarium spp. *Archives of Microbiology,* Springer, ISSN: 1432-072X; 0302-8933; https://doi.org/10.1007/s00203-021-02520-2.

2020. "Nitrate Bioremoval by Phytotechnology using U aurea Collected from Eutrophic Lake of Theerthamkara, Kerala, India," Polln. ISSN- 2383-4501. Vol. 6(1):149-157.

2020. "Combined effect of Nitrate Bioremoval by aquatic free-floating plant an association with filamentous Cyanobacteria," *Austin Environ Sci* 5 (1), 1-4.

2018. Nitrate removal by aquatic free-floating plant-based bioremediation from eutrophic lake of Theerthamkara India. Book of Abstracts, ISBN: 978-618-81537-6-9*, e-proceedings of the 7th European Bioremediation Conference*, Technical University of Crete, Greece.

2017. Diesel biotreatment competence of indigenous methylparathion degrading bacterial strain of Pseudomonas aeruginosa mpd. *Journal of Microbiology, Biotechnology Food Sciences.,* JMBFS-ISSN-1338-5178; 6(3):878-885.

2017. "Determination of nitrate utilization efficiency of selective strain of bacillus sp isolated from Eutrophic Lake, Theerthamkara, Kasaragod, Kerala." Polln. ISSN- 2383-4501. Vol. 3(1):55-67.

2016. Box-Behnken experimental design mediated optimization of aqueous methylparathion biodegradation by Pseudomonas aeruginosa mpd-5 strain. *Journal of Microbiology, Biotechnology Food Sciences.*, ISSN-1338-5178; 5(6): 534-547.

2016. Aqueous methylparathion removal by Ozonation and Optimization of variables using Central Composite Design of experiments. *Current Environmental Engineering*, ISSN-2212-7186; 3(3): 245-266, Bentham Science Publishers.

2016. "Diesel oil utilization efficiency of selective Bacterial isolates from Automobile workshop and Thesjaswini river of Kerala." Polln. ISSN-2383-4501. Vol. 2(2): 221-232, Spring.

2015. "Combined Mycobiotreatment and Ozonationt Process for the efficient removal of toxic methylparathion from Wastewater." Book of Abstracts, ISBN: 978-960-8475-23-6, Grafima Publications *e-proceedings of the 6th European Bioremediation Conference*, Technical University of Crete, Greece.

2013. Optimization of Phosphate removal from synthetic wastewater by bacterial consortium using Box-Behnken Design. *Environmental Engineering and Management Journal.*, 2013, Vol. 12(12), pp: 2371-2383.

2013. Optimization of aqueous methylparathion biodegradation by Fusarium sp in batch scale process using response surface methodology. *International Journal of Environmental Science and Technology.*, 2013, Vol. 10(3), pp: 591-606., Springer Publications.

2012. Effect of pH on the degradation of aqueous organophosphate (methylparathion) in wastewater by ozonation. *International Journal of Environmental Research.*, ISSN: 1735-6865, 2012, Vol. 6(2), pp: 557-564, Springer. Bioline International.

2011. Biological Removal of Phosphate from Synthetic Wastewater Using Bacterial Consortium. *Iranian Journal of Biotechnol.* 2011, 9 (1), 14. National Institute of Genetic Engineering and Biotechnology, Iran.

2010. Studies on the Removal Efficiency of Phosphate from Wastewater using Pseudomonas sp YLW-7 and Enterobacter sp KLW-2. *Glob. J. Environ. Res.*, 4 (2), 83-89.

2010. Physico-chemical and Bacteriological characteristics of Noyyal River and groundwater quality of Perur, India, *J. Appl. Sci. Environ. Manage., Bioline*, 14 (2), 29-35.

2010. Bio-treatment of phosphate from synthetic wastewater using Pseudomonas sp YLW-7. *J. Appl. Sci. Environ. Manage., Bioline*, 14 (2), 75-80.

2010. "Laboratory application of Ozonation process in the treatment of organophosphate contaminated wastewater." *Proceedings of 3rd International Conference on Hydrology and Watershed Management,* Organized by Centre for Water Resources, Institute of Science and Technology, Jawaharlal Nehru Technological University, Hyderabad (AP), India. BSP Publications, www.bspublications.net, ISBN: 978-81-7800-224-8., VOL. II. Pp-1221-1231. Copyright © 2010.

2009. Studies on the efficiency of the removal of phosphate using bacterial consortium for the biotreatment of phosphate wastewater. *Eur. J. Appl. Sci.*, 1, 06-15.

Professor. Dr. Gustavo Henrique Ribeiro Da Silva

Associate Professor,
Department of Civil and Environmental Engineering,
UNESP, Sao Paulo State University, Brazil
Email: gustavo.ribeiro@unessp.br

Professor, Dr. Gustavo Henrique Ribeiro Da Silva, Associate Professor, Department of Civil and Environmental Engineering, School of Engineering, UNESP, Sao Paulo State University, Brazil. Since obtaining his master's degree, he has extensive experience with water treatment research including wastewater treatment by microalgae, nutrient removal, and sustainable wastewater management by constructive wetlands, all areas of research with the utmost environmentally friendly technologies. He is a person of high achievement in remediation work, an area of critical importance. He is scientifically knowledgeable in environmental engineering, ozonation, and environmental technology of wastewater. Moreover, he worked in the treatment of water and wastewater using microalgae with sustainable technology. He has a plethora of scientific publications including more than 50 research scientific papers with well-reputed scientifically peer-reviewed journals, more than five book chapters, for well-reputed and scientifically recognized peer-reviewed international journals and publishers, more than 96 papers for international and national conferences to his credit. He has more than 13 years of teaching and 15 years of research experience and acts as a referee and reviewer for scientific reputed journals and publishers. He supervised several PG, PhD., and Post-Doctoral students both at the national and international levels and has also served on the board of members for scientific committees, academic, research, and industries in several organizations, institutions, and forums. He has ten projects and grants to his credits, collaborated with national and international scientific collaborators, and also acted as a scientific project referee for more than five countries

Orcid: *https://orcid.org/0000-0002-0741-8966*
Google Scholar:
https://scholar.google.co.in/citations?user=cfoG94kAAAAJ&hl=en
SCOPUS ID:
https://www.scopus.com/authid/detail.uri?authorId=55337052000
Researchghate: *www.researchgate.net/profile/Gustavo-Da-Silva-3*
CNPq, Brazil: *http://lattes.cnpq.br/9415339655312598*

Index

E

F

G

H

I

K

L

M

S

T

W

X